THE
OXFORD BOOK OF
BRITISH BIRD NAMES

The Oxford Book of
British Bird Names

W. B. LOCKWOOD

Oxford   New York

OXFORD UNIVERSITY PRESS

1984

Oxford University Press, Walton Street, Oxford OX2 6DP

London New York Toronto
Delhi Bombay Calcutta Madras Karachi
Kuala Lumpur Singapore Hong Kong Tokyo
Nairobi Dar es Salaam Cape Town
Melbourne Auckland
and associated companies in
Beirut Berlin Ibadan Mexico City Nicosia

106685

Oxford is a trade mark of Oxford University Press

British Library Cataloguing in Publication Data

Lockwood, W. B.
The Oxford book of British bird names.
1. Birds—Great Britain—Nomenclature
I. Title
598.2941   QL677
ISBN 0–19–214155–4

Library of Congress Cataloging in Publication Data

Lockwood, W. B. (William Burley)
The Oxford book of British bird names.
Bibliography: p.
1. Birds—Great Britain—Nomenclature (Popular)—Dictionaries. I. Title
QL690.G7L593 1984   598.2941′014   84–4362
ISBN 0–19–214155–4

Phototypeset by Wyvern Typesetting Ltd, Bristol
Printed in Great Britain by Billings & Sons Ltd,
Worcester

FOR PRAVIN AND VERA

initiators – unwittingly – of the
present undertaking.

FOREWORD

THE present widespread and, by all accounts, growing interest in birdlife is naturally accompanied by a certain curiosity about the names we give our birds, for they are often so puzzling. Many must have wondered how a bird came to be called a Hobby or a Rail, and what is one to make of Titmouse or Wheatear? Local terms may be no less odd, like Elk for Whooper Swan or Olive for Oystercatcher, or the equally bizarre Cherry Sucker, Heather Bleat, Sandy Loo for the Spotted Flycatcher, Snipe, and Ringed Plover respectively, to mention just these. Nightingale and Nightjar have plausibly something to do with the night, but what of the second part of these names? It is the same with Nuthatch and Redstart. Even when the name is entirely transparent, a problem may still remain. Why, for instance, is the Blackbird so called, seeing that other species, as the Carrion Crow and the Raven, are quite as black?

Writers on ornithological subjects sometimes digress to insert remarks here and there on the meanings of obscure bird names. But, all too commonly, the speculation volunteered is worthless. It is evidently still not everywhere appreciated that the elucidation of bird names is as much an art in its own right as any other branch of linguistics, one which necessarily calls for specialist philological training. Only insight gained through such study could reveal that Thrush originated, some three thousand years ago, as a variant of the ancestral form of Throstle. The interpretation of a name may have consequences for scientific ornithology too, witness Fulmar or Woodwall. Aspects of folklore are not infrequently explicable when the hidden history of a name is uncovered by philological means. The results may indeed be startling. Having discovered that the basic meaning of Stork is stick, it was but a short step to understanding at last where the notion of the baby-bringer comes from.

Admittedly, not all the questions can as yet be answered with certainty, and there is good reason to believe that in a number of cases answers may for ever elude us. Nevertheless, great advances have been made in clarifying the origins and development of names, and the present publication aims to supply the conclusions available to date. Here too, truth may be stranger than fiction. Who would have dreamed that Swallow and Solan are basically the same word or that Fieldfare has every appearance of being a corruption of an Old English name meaning grey piglet?

The Introduction considers the principles involved in the interpretation of bird names. The main part of the work is arranged as a dictionary.

It includes both standard and local names, with variants amounting to some 1500 items relating to 257 species, thus covering all our commoner birds plus a fair sample of the less usual visitors. All the modern standard names are listed, but, in the face of overwhelming numbers, the entries could not be exhaustive in respect of other names. As a selection was inevitable, it was important that it should be representative, the intention being to exemplify all the name types associated with a given species. In this way, a reader seeking information on a name or variant not quoted here, should have little difficulty in finding, by proxy, at least some general guidance.

W. B. LOCKWOOD

Department of German
Reading University
September 1983

CONTENTS

INTRODUCTION

THE story of our language goes back to the middle of the fifth century, when Angles and Saxons, leaving their settlements in the coastlands of the Low Countries and north-west Germany, began the conquest of Britain. By the year 600, or at the latest by 650, the invaders had seized all the land south of the Scottish border except the far north-west (roughly Cumberland and Westmorland), Wales, and the Devonian peninsula; north of the border, their conquests took in the eastern lowlands up to the Firth of Forth. The language of the conquerors is known as Anglo-Saxon or Old English; we shall use the latter term. By the seventh century then, this language had established itself over most of the country, the Celtic speech of the earlier inhabitants, Old British as it is called, dying out in the parts taken over by the newcomers. Latin had been introduced to Britain during the Roman occupation from 43 to about 410 AD and was doubtless still being spoken here, especially in the Romanized urban areas, when the Angles and Saxons arrived. This language now also died out.

The Angles and Saxons were not to remain in possession undisturbed. From the late eighth century the country suffered almost continuously from the depredations of Viking raiders who, from 876 onwards, began to settle here permanently and soon held a large part of the north and east. This territory became known as the Danelaw, where for several centuries the language of the invaders, Old Norse, was spoken, in some places exclusively, in others side by side with the native English. Also in the ninth century Norwegian Vikings occupied Cumberland and Westmorland. The political independence of these foreigners came to an end in 954, but Scandinavian influence continued to be a factor in the national life; indeed, in the early part of the eleventh century Danish kings sat upon the throne of England. How long Old Norse maintained itself here is uncertain, but it probably survived in certain districts until the end of the twelfth century. As a Germanic language, Old Norse was rather close to Old English.

The Anglo-Saxon period came to an abrupt end with the warlike arrival of the Normans in 1066. These conquerors spoke a northern form of Old French, which remained the medium of the ruling circles and their entourage for some two or three centuries. With the advent of the Normans, the Old English literary language quickly fell into disuse, though of course English continued to be spoken by the vast majority of the population. It was not until the thirteenth and fourteenth centuries that English came into its own again as a literary language. By then, however, it had become very different from the language of Anglo-Saxon times, and much more like the English of today. It is called Middle English, and this term is employed to denote the language until the beginning of modern times, c.1500. The language after this date is defined as Modern English.

English has greatly extended its geographical range since the Anglo-Saxon period. It has largely replaced the Celtic remaining south of the Scottish border, only Welsh now surviving in parts of Wales, although Cornish lingered on in west Cornwall until the end of the eighteenth century. English now dominates in Scotland and Ireland where, until the beginning of modern times, Gaelic—another form of Celtic—was the most widespread language. Today, Gaelic is confined to small areas only, in Scotland chiefly to the Hebrides, in Ireland to a few districts in the west. The Gaelic of the Isle of Man has been extinct since the middle of the present century. Irish, Scottish, and Manx Gaelic are in practice so diverse as to constitute separate languages, even though the expression Gaelic is conventionally applied to all of them. During the Viking age, Old Norse also established itself in various parts of the then Gaelic-speaking countries, as follows: at certain points on the Irish seaboard, along the west coast of Scotland and notably in the Hebrides, and on the Isle of Man. In these places Old Norse probably died out, or was at least moribund, by the end of the twelfth century—more or less as in England—and then Gaelic surged back. In the far north, in the Orkney and Shetland Islands, Old Norse quite replaced the ancient indigenous language, Pictish. Here the Norse language became known as Norn. Orkney and Shetland Norn remained in use until the first half of the eighteenth and nineteenth centuries respectively, by which time these dialects had finally succumbed to the English introduced from Scotland.

It is apparent that, throughout its history, English has been in direct contact with other languages spoken in these islands. To a greater or lesser degree such contacts have influenced English and explain why many of our bird names originally came from these languages; in this respect French has been particularly significant. On the nature of these contacts, see p. 11.

Dialect and Standard Language

The invading Angles and Saxons were illiterate. Their language, a purely oral medium therefore, would in the nature of the case fall into different dialects. Writing in English appears to have begun towards 700, necessarily on a dialect basis. Four dialects can be distinguished: Kentish and West Saxon covering the area south of the Thames, Mercian in the central part of the country and, north of the Humber, Northumbrian. The earliest documents are eighth-century glossaries which include several bird names, but most of the records of Old English date from the tenth and eleventh centuries. They are written chiefly in the West Saxon dialect.

Middle English, too, was at first written in one or other of the regional dialects, but that of London was the most significant and here the Mercian style was paramount. It found its highest expression in the works of Geoffrey Chaucer (died 1400) and by the end of the fifteenth century this type of English had become the accepted written norm. The introduction of printing in 1476 had resulted in a great increase in literacy and gave the written language an unprecedented importance, quickly spreading its use throughout the country. Furthermore, the written word now became the model for the speech of the educated everywhere. The standardization of the written word led to the beginning of the standardization of the spoken word too. The regional dialects, however, continued to be used by ordinary, less educated people until the Industrial Revolution brought a great part of the rural community together in the towns and cities. In these new urban centres, with their mixed populations, there was no place for broad dialect; they became strongholds of the spoken standard. The dialects in the countryside could maintain themselves a little while longer, but were increasingly modified in the direction of Standard English as literacy became universal in the latter half of the last century and modern communications broke down the isolation of remote districts. By the early years of this century, pure dialect speech in England—though not yet in Scotland—was approaching vanishing point. Today, three generations later, all that remains is the local flavour as a reminder of what once was. The Modern English period may therefore be characterized as the period during which the language achieved its present high degree of uniformity, first in its written, then in its spoken form.

These developments are duly reflected in the use of our bird names. Where a given bird had more than one name, and this applied to most species, the concept of a standard term would not, and could not, emerge until the Modern English period. At the same time, as long as the dialects flourished, the employment of dialectal bird names, which as much as any category of names were identified with the countryside, continued unabated. Many survive to this day, helping to impart to rural speech that local flavour we have referred to. But since dialect speaking proper is now extinct in England, we have preferred to call such names regional or local, rather than dialectal. The information available on the distribution of these names refers in the main to the eighteenth and nineteenth centuries; no comparable records exist for the present century. In this respect, therefore, our work perforce reflects the linguistic situation before standard usage had made itself so strongly felt.

Linguistic Evolution

Living languages are constantly changing. It is not difficult to see how this applies to the vocabulary: new words come into being, old ones may be forgotten, as can easily be illustrated from the bird-name stock. Of the seventy or so bird names found in the Old English sources, only some thirty remain today. The places of the others have been taken by names of more recent origin, some arising spontaneously within English, others coming in from foreign languages.

The phonetic structure of words has likewise undergone change. In many cases, three phonetic stages can be distinguished, corresponding to the main historical periods, i.e. Old, Middle, and Modern English. Thus Modern English *rook* goes back via Middle English *rōc* to Old English *hrōc*. Sometimes only two different stages are apparent: Modern English *crow*, Middle and Old English *crāwe*, or Modern and Middle English *finch*, Old English *finc*. All these changes are subject to definite rules, or phonetic laws, as they are called. A problem may, however, occasionally arise in connection with the Old English forms. The surviving records of Old English are, for the greater part, written in the West Saxon dialect, but it was on the basis of the rather poorly attested Mercian dialect that Standard English evolved. A bird name in West Saxon may well have had a somewhat different form from its (very probably unrecorded) Mercian equivalent. It is seen that *sparrow* goes back to Middle English *sparwe* according to rule, a form which implies not the actually recorded West Saxon *spearwa*, but an unattested Mercian *sparwa*—an asterisk is used in these studies to indicate a word theoretically reconstructed, but not actually recorded.

Our oldest bird names were, of course, already present in the language of the Angles and Saxons. Whereas many of them would actually be created in this country during the Old English period, others were certainly in existence at the time of the invasions. Such names must have been brought from the Continent, where the prehistoric roots of the English language are to be found. The speech of the continental Angles and Saxons was an ancient form of German. It was closely allied to another ancient form from which the present-day German language is descended, and even more closely to a form which was the ancestor of Dutch. This explains why one often recognizes English bird names again in German or Dutch. It will be realized that each of these languages too, has been continuously evolving, the shape of the words changing over the centuries. Thus English *swan*, Dutch *zwaan*, and German *Schwan* are reflexes of the separate evolution in the three different languages of an originally identical word, and the same process is seen again e.g. in English *swallow*, Dutch *zwaluw*, German *Schwalbe*.

In order to appreciate linguistic aspects fully, it will in certain cases be important to observe that German, as normally understood today, is technically High German, that is to say, it was in the first place the language of the centre and south of the country only. As such it varies noticeably from the indigenous language of the north, technically Low German, now sunk to the level of a patois. Dutch is much closer to the Low German type than to High German. Mention must also be made of the minor language Frisian, close to Dutch and Low German, spoken in Friesland, Netherlands (West Frisian) and on the west coast of Schleswig, Germany, and adjacent islands (North Frisian), to which we need to refer in the case of certain bird names. Naturally enough, historical periods can be distinguished in these languages, e.g. Old High German, Middle High German, corresponding to Old English, Middle English.

The earliest forms of continental German were spoken dialects only, the language at this stage having no written literature. This language, the common ancestor of English, Dutch, and German, is termed West Germanic. It is to be envisaged as existing during, roughly, the first three centuries AD. It follows that where originally identical words are found in English, Dutch, and German, as in the case of the Swan and Swallow names above, the prototypes will have existed in West Germanic. Or, expressed in another way, it could be said that English *swan*, Dutch *zwaan*, German *Schwan*, etc., are of West Germanic age.

West Germanic itself was, in turn, related to North Germanic, the contemporary language of Scandinavia, likewise a spoken language only. North Germanic is the parent of Old Norse, the language of the Viking age, c.800–1100, and well known from saga literature. Old Norse is the common ancestor of present-day Swedish, Danish, and Norwegian. As we have seen, this language was brought to the Scandinavian settlements in the British Isles, though it eventually died out there. It was also carried to the Faroes and Iceland, where it developed into the modern languages, Faroese and Icelandic.

In the centuries before the birth of Christ, West and North Germanic had not yet become markedly differentiated and were still so similar that they are to be regarded as variant forms of a single language. This earlier language is known as Proto-Germanic; needless to say, it was unwritten. When it arose is uncertain, but one may think in terms of 500 to 1000 BC. Now it occasionally happens that bird names of West Germanic age have parallels in Old Norse, as indeed is the case with the words for Swan and

Swallow: *svanr, svala.* One concludes that such names will be traditional in Scandinavia and of North Germanic age. One may further reason that, since these are then common to both North and West Germanic, they must have been present in the preceding Proto-Germanic as well. By the application of phonetic laws, philologists can often postulate the form a given word will have had even at that remote date. It can be taken as certain that the Proto-Germanic forms of the two bird names just mentioned were **swanaz* and **swalwō-*, where *swan-* and *swalw-* are described as roots, *-az* and *-ō* as inflexional endings.

We do not wish to imply that only names actually attested in both the West and North divisions of Germanic can be considered as of Proto-Germanic age. That is not so, for it can happen that a name found, for instance, only in West Germanic, as Ouzel, can be shown philologically to have been in existence in Proto-Germanic. If such names are found solely in West Germanic, that must be due to losses in North Germanic, and vice versa.

The Proto-Germanic language itself is a branch of a large family of languages, the so-called Indo-European family. Those closest to it are Proto-Italic and Proto-Celtic, likewise prehistoric languages of comparable antiquity, more or less, the forms of which can also be, in large measure, reconstructed theoretically. By far the most important ancient Italic language is Latin, from which French, Italian, Spanish, and the other Romance languages are derived. It may be added that Standard French is of northern origin, being based on the dialect of the Paris region. In the south of the country, the local Provençal, today only a patois, is in reality an independent Romance language. The philologically most significant Celtic languages are Irish Gaelic and Welsh; in antiquity, however, Celtic was widely spoken on the Continent, but died out there without leaving any literary records. Other branches of the Indo-European family are Proto-Slavonic, the ancestor of the Slavonic languages, as Russian and Polish, and Proto-Baltic, the most important descendant here being Lithuanian, and further Proto-Hellenic, from which comes Greek. Speakers of some languages of this family found their way to the east, in particular we notice those who finally settled in India. Their language is suitably named Proto-Indic, and from it originated Sanskrit, the classical language of India.

All these multifarious tongues stem from a common source called Proto-Indo-European , a language spoken in Central or Eastern Europe in the first half of the third millennium BC. In the same way as philologists are able to restore theoretically the forms of Proto-Germanic and the rest, so can they determine, at least in outline, the forms of the parent Indo-European. It is noticeable that not all branches of the family are of the same age. Of those mentioned above, Proto-Indic and

Proto-Hellenic separated from the main body during the third millennium, while the others can hardly have emerged as distinct entities until towards the end of the second millennium. Furthermore, these others remained in close geographical proximity, which implies a degree of continuing evolution in common.

We have seen how, by a process of comparison, certain names can be traced back to Proto-Germanic. By a similar process, it is in principle possible to trace a name further back to pre-Germanic times, i.e. into the so-called Indo-European period. However, not more than a handful of our bird names are so ancient that they can be shown to have existed at such a remote date. Moreover, since bird names are morphologically among the less stable items in the lexicon, the very ancient names are not—with one exception—traceable to a single Indo-European form. At best, one can manage to identify related variants, see e.g. STARLING or THROSTLE. Solely in the case of Goose can an original form be posited, giving it a strong claim to be considered our most archaic bird name.

Germanic Consonant Changes

The differences between the various branches of the Indo-European family are naturally more fundamental than the differences which have developed between the separate languages within each branch. For our immediate purpose it will suffice to observe that certain consonant changes clearly mark off the languages of the Germanic branch from those of the other branches. Thus English *three* and its Germanic cognates (as Dutch *drie* , German *drei*, but in both languages at the early medieval stage *thrī* Old Norse *þrīr*) correspond to Latin *trēs*, Greek *treîs*, Sanskrit *trāyas*, Irish *trī*, Welsh *tri*, Lithuanian *trys*, Russian *tri*. In this etymological equation one notes that the Germanic languages have, or originally had, *th* where the others have *t*. It is evident that the Germanic has innovated, the other languages preserving unchanged the typical Indo-European consonant. In the same way, Germanic shows *k* from Indo-European *g*; compare English *knee* (*k* formerly pronounced, as still in Dutch *knie*, German *Knie*; Old Norse *knē*) with synonymous Latin *genu* or Greek *gónu*. Similarly Germanic *f* and *t* answer to Indo-European *p* and *d* respectively: English *foot* (Dutch *voet*, German *Fuss*, both ultimately from *fōt*; Old Norse *fōtr*) as against Latin (root) *ped-* or Greek *pod-*. A final example: English *heart* (Dutch *haart*, German *Herz*, both ultimately from *hert*; Old Norse *hjarta*) cognate with Latin (root) *cord-*, Greek *kard-*, where *h* in the Germanic languages has arisen from Indo-European *k*, and again Germanic *t* answers to Indo-European *d*. The above changes, forming part of the so-called Germanic Sound Shift, may be tabulated:

Indo-European consonant *t* becomes Germanic *th*

k   *h*

p   *f*

d   *t*

g   *k*

and similarly *zd* becomes *st*; see THROSTLE.

To the above rules there are occasional exceptions. Suffice it here to note that Indo-European *p* and *t* in the combinations *sp*, *st* remain unchanged in Germanic (see SPINK, STARLING) while Germanic *f* and *h* can sometimes be further shifted to *v* and fricative *g* respectively (see RAVEN, HERON). The Sound Shift cannot be dated precisely, though it certainly took place in prehistoric times. Philologists are inclined to think in terms of between 500 and 250 BC, so that a name which can be shown to have passed through the Shift must have been in existence before then.

It should be added that in certain cases a non-Germanic branch too, may change an original Indo-European consonant in accordance with some phonetic law peculiar to that branch. Thus the Sanskrit word for knee is *jấnu*, i.e. with *j* developed from original *g* seen in Latin *genu* and Greek *gónu*. At times these changes can be very complex and affect several branches, see e.g. GOOSE. Since these, like other linguistic matters touched upon above, cannot be fully treated here, the interested reader may be referred to the author's *Indo-European Philology*, 1977.

The Germanic Sound Shift is of special relevance where old bird names are of onomatopoeic origin. We begin by applying the law of consonant changes to find the origin of *finch*. The Old English form was *finc* and we recognize that the name is certainly of West Germanic age, since it occurs in Dutch *vink* and German *Fink*. But as it stands, its ulterior connections are entirely obscure. Perhaps philology can help. Now a name of West Germanic age may well have existed in Proto-Germanic and could therefore possibly have been present in pre-Germanic times as well. We shall, for the sake of argument, assume that this was in fact the case. This being so, the Proto-Germanic root **fink-* presupposes Indo-European **ping-* (*g* being sounded). At once we get a root reminiscent of *pink*, a common English dialect name for the Chaffinch, an imitation of the bird's unmistakable call. The postulated Indo-European **ping-* must be essentially the same thing. In other words, some three thousand years ago, in that area of Indo-European speech from which Germanic emerged, there already existed a simple, echoic name for the Chaffinch, for this must be the species in question. We have thus not only a satisfactory explanation of the origin of the name *finch*, but also proof that it was first applied to the Chaffinch, with the rider that its use to denote other species, as Bullfinch or Greenfinch, must be a secondary development. It is not without interest to observe that the originally onomatopoeic character of the name was much weakened when it passed through the consonant shift. This left the way open for the creation, or recreation, of a comparable synonym *pink*; to that extent history has repeated itself.

It was noted above that *rook* goes back to Old English *hrōc*, where of course the *h* was clearly sounded. This name has regular correspondences in Old High German *hruoh* and Old Norse *hrōkr*. One can therefore confidently postulate a Proto-Germanic root **hrōk-* which in turn could imply Indo-European **krōg-*, or (in accord with the laws of vowel change) alternatively **krāg-*. Indeed, this must have been the case, since the Indo-European roots thus postulated at once suggest an onomatopoeic origin, an imitation of the deep, grating cry of the Rook and other members of the Crow family. Just as with *finch*, the basic onomatopoeia of the Indo-European stage is obscured by the working of the Germanic Sound Shift, to be totally lost in the modern form *rook* owing to phonetic developments since Old English times.

The philological analysis of *raven* leads to comparable results. The name goes back to Old English *hræfn*, with regular correspondences in Old High German *hraban* and Old Norse *hrafn*. These forms presuppose Proto-Germanic **hravnaz*, where *-n-* and *-az-* are inflexional elements added to a root **hrav-*, originally **hraf-*. The consonants here imply Indo-European *k*, *r*, *p*. Philological considerations indicate that, in this case, the Proto-Germanic root vowel *a* descends, as it often does, from Indo-European *o*, seen e.g. in Latin *corvus* raven, further Greek *kórax*. We therefore posit **krop-* as the ultimate root, and observe again a formation explicable in terms of onomatopoeia.

We next turn to the semantic aspects. There are no grounds for doubting that the Proto-Germanic **hraf-* properly denoted the Raven, but the meaning of **hrōk-* cannot have been so specific. It is recognized that the name was inspired by the voice; its meaning must then have included the two Crows (the Carrion and the Hooded) whose cries are so similar to the Rook's, and the species are of course comparable in size and appearance. In this connection one observes that Latin *corvus* and *cornix*, like Greek *kórax* and *korṓnē*, conventionally translated raven and crow respectively, had the same semantic range, the former being properly restricted to the Raven, the latter covering in practice Crow and Rook as well. This was, so it appears, the ancient way of doing things. One observes, too, that the Latin and Greek names are built out of echoic elements not unlike those underlying the Germanic.

The cawing of both Rook and Crow is so insistent that it comes as no surprise to learn that names for these species based on the call are universal. In Proto-Germanic, however, the echoic origin of their name was no longer fully apparent, having been blurred by the Sound Shift. It is not hard to imagine that there would then be a need for new name which could be felt as representing the so familiar cry. It was out of this need that there arose the ancestral form of English *crow* and its cognates Dutch *kraai* and German *Krähe*, evidently of West Germanic age. In North Germanic its place is taken by Old Norse *krāka*. Both types are patently imitative. It is equally apparent that both were created in a Germanic context. They are thus more recent than **hrōk-*, for which we suppose them to have been, in the first place, popular synonyms with a comparable range of meaning, i.e. covering the Rook and the Crows together. At the same time, thanks to the presence of distinctive doublets, the evolving Germanic languages now had the resources to distinguish readily between the Rook and the Crows. It was simply a matter of narrowing down the meaning of the two synonyms: one had to be restricted to *Corvus frugilegus*, the other to *C. corone corone* and *C. c. cornix*. Precisely this took place, hence the specific meanings of English *rook* and *crow*, paralleled in Dutch *roek* and *kraai*. Modern German has lost its Rook cognate, but it was present in medieval German (see *hruoh* above); the current term *Saatkrähe*, literally seed crow, is, of course, a much later creation. A similar situation doubtless obtained between Old Norse *hrōkr* and *krāka*. Incidentally, it may be noticed that the Germanic languages were thus able to settle a terminological problem which remained unsolved, at this level, in Latin or Greek.

Motivation and Changes in Bird Names

Our traditional bird names arose in a purely oral milieu. They were created by country folk, in the case of sea birds sometimes by fishermen, in times when people still lived on intimate terms with nature. The names have been variously motivated. Habits and appearance, especially colour, have been predisposing factors. But the most potent of all has been the voice—so many species are heard before they are seen—hence the plethora of onomatopoeic (imitative, echoic) names, in which the bird's call, or one of its calls, is reproduced as the name. Since a bird's cry is physiologically not at all the same thing as a human utterance, the cry cannot be truly expressed in this way. Onomatopoeic representation is thus impressionistic, subjective, however realistic we may unconsciously suppose it to be. It is not difficult to test this. Pigeons are said to coo, and we are all convinced that they do, in fact, call 'coo'. It doesn't easily occur to us that they might be calling 'doove', but that is how our ancestors heard it, hence the name DOVE. For another imitative rendering, see TURTLE. In this way, quite disparate echoic names for one and the same species may come into existence. Not infrequently the sounds made by birds are interpreted, in whole or in part, as meaningful, though usually nonsensical, words, e.g. CHIFFCHAFF or the regional name BUTTER BUMP for Bittern. Finally, it may be noted that traditional names often have little in common with scientific nomenclature, which may be based upon considerations hidden from the layman's eye. Thus the Jay is never classed as a Crow in popular terminologies.

Bird names are prone to change, in the sense that old names often drop out of use, while new names come into the language. The oldest names are simplexes, like RUDDOCK for Robin, the only form occurring in Old English. Compounds like REDBREAST arose later, ROBIN REDBREAST later still, the use of a Christian name for a bird being unknown in the earlier period. Sometimes precise reasons for the changes can be given. The working of superstition, for example, has caused some old names to fall under taboo and to be replaced by harmless, evasive substitutes, the so-called noa names, as in several local terms for the Swift; see DEVIL 2. A particularly potent factor for change has been the operation of folk (or popular) etymology. By this is meant the tendency to modify a name which in its form is unusual, or otherwise appears peculiar, by associating it with some familiar word. The results may be bizarre, as when original YELLOW AMMER was 'corrected' to YELLOWHAMMER. YAFFLE, a local name for the Green Woodpecker, became in some districts YAFFINGALE after the analogy of NIGHTINGALE. The complicated history of LAPWING is nothing more than a series of such shifts. Altogether folk etymology has been one of the mainsprings in the unceasing proliferation of names so characteristic of the language until the eighteenth century, after which the decline of the rural communities and the spread of standard terms virtually put an end to the spontaneous rise of new names.

How old are our Bird Names?

Only in the case of names expressly coined by ornithologists or other specialists (see below under 'Standardization of Names'), is it possible to date precisely the creation of a name. But the great majority of our bird names are folk names, arising anonymously. One can, of course, find the date of the first recording of such names. Important as such dating is, it cannot tell us how long the name had been in use previously. Indeed, the first recording may be fortuitous enough. For example, SHELDRAKE is not noticed in an ornithological context prior to c.1325; as a surname, however, the record takes us back to 1195. DOVE is absent from the Old English sources which use other names, as *culfer* or *turtla*, yet it must have been present from the earliest times, witness its counterparts in Dutch *duif* and German *Taube*.

It goes without saying that French bird names are not likely to have entered the English language before 1066, but would have been adopted after the Conquest during the period of Norman ascendency, when French was the language of prestige in this country, from which the native English borrowed avidly. Where a name has been taken from Norse, the general historical background again supplies at least an approximate chronology. Norse, however, unlike Norman French, never attained more than local, or at best regional, significance. It will therefore not be accidental that bird names of Scandinavian provenance are not found in the Old English records, but first appear in the Middle English period. In other words, such Norse bird names as passed into English did so only after Norse had given way to English. When a population gives up its own language in favour of another, certain specialized parts of the vocabulary of the original tongue tend to persist and be carried over into the new language, as spoken locally. In this respect, bird names can be most tenacious, a propensity which naturally also explains the presence of numerous Cornish, i.e. Cornish Celtic, bird names still in use in Cornwall or, no less strikingly, the medley of bird names of Norn origin to be heard in the speech of the people in Orkney and Shetland. Similarly, in the anglicized parts of Wales and Scotland, the influence of the former indigenous languages is seen in certain bird names. In all these places too, the old names would enter the incoming English as the native languages died out. It follows, incidentally, that words from these lost languages tend to remain local. Only occasionally have they become general, and then for some special reason; see e.g. AUK (Old Norse), BONXIE (Norn), GULL (Cornish), PTARMIGAN (Scottish Gaelic).

Apart from occasional exceptions, e.g. CULVER from Latin, all names going back to Old English times are clearly native Germanic, even though a considerable number are not actually recorded in the other Germanic languages. To some extent this may be fortuitous, but in the nature of the case there can be little doubt that the majority of the names in question were spontaneously created in this country and cannot therefore be older than the Anglo-Saxon period. The demonstrably oldest names are naturally those which have congeners in other Germanic languages, for these must have been in existence before the invaders left the Continent. Among these, one can again recognize chronological strata, as explained on pp. 5–7.

It remains to be said that several native Germanic names in use in this country entered our language via French, e.g. HERON. The reason for this, at first sight, paradoxical situation, is as follows. In the fifth century,

Gaul was invaded by the German-speaking Franks—it was these who gave Gaul its present name of France. The newcomers occupied the northern part of the country in strength, for long employing their own language. As a result, many German words passed into French, among them an appreciable number of bird names, some of which later reached us through the Normans.

Standardization of Names

The general evolutionary trend towards uniformity in the language naturally paved the way for the highly standardized ornithological nomenclature current in English today. In this connection we must also consider those bird names which have not arisen as folk names, but which have been introduced by scholars. These form an appreciable contingent; in fact quite one half of the standard names of the species referred to in this book, including some very common ones, are of this sort.

The origins of the standard nomenclature can be traced back to the writings of the earliest ornithologists, beginning with Turner 1544—for bibliographical details of this and other works, see pp. 17–20. Their primary sources were the natural histories of Pliny and Aristotle, the latter best known through Gaza's Latin translation of 1476. Of medieval writers, Albertus Magnus c.1250 was by far the most significant, until the naturalists of the sixteenth century inaugurated the modern era, foremost among them Gesner 1555. The language of this literature was Latin, and Latin was still the common language of scientific work in Linnaeus' day, two centuries later—our present-day scientific terms are a last reminder of the tradition of learned writing in Latin.

From the second half of the seventeenth century, bird books begin to appear in English too: Merrett 1667, Charleton 1668, and, above all, Ray 1678. In these works we find several names which are not traditional English, but translations from the Latin of earlier authors. Thus in Ray 1678 occurs the name MOUNTAIN FINCH which, as a term for the Brambling, was taken up by some later authors; it is a rendering of Latin *Montifringilla*. Many of these neologisms, like the one just mentioned, have since passed out of use, but others remain and may have replaced traditional names. Outstanding examples are seen among Tit names, e.g. Ray 1678, Blue Titmouse, Marsh Titmouse translating *Parus caeruleus* and *P. palustris* respectively. It may be further noted that several exotic names culled from the works of foreign ornithologists have gained acceptance, even though genuine English names have been available, as AVOCET or GARGANEY.

The final standardization of names was, however, essentially due to precedents set by later writers. Pioneering, and the most influential in this respect, was Pennant, whose *British Zoology (Birds)*, 1768, began the practice of adapting English names, or coining new ones, to accord with the scientific classification initiated by Linnaeus 1758; for examples, see BUNTING, whence REED BUNTING, etc. Pennant's successors continued to develop names in this way to keep pace with advances in systematics, though not always agreeing among themselves as to what was most appropriate. But from Yarrell 1843 onwards, a high degree of standardization is evident, the tendencies to uniformity culminating in *A List of British Birds* compiled by a Committee of the British Ornithologists' Union, 1883. Here one vernacular name only was admitted for each species; where Yarrell had, for instance, LAND RAIL or CORNCRAKE, the new prescription was Corncrake. Since then, only a few minor changes have taken place, e.g. WOOD PIGEON for RING DOVE, from Hartert 1912.

We do not wish to imply that terminology in English is now as consistent as the scientific nomenclature in Latin. Despite all refinements, numerous terminological inconsistences remain and some are likely to continue; see e.g. SPARROW. Although a standard nomenclature in English now exists, it is not of course exclusive, for ordinary users of the language do not necessarily feel bound by the prescriptions of the ornithologists, indeed they will generally not even be aware of them. The standard name ROBIN, a most suitable term from the point of view of systematics, may very well be called REDBREAST, or for that matter ROBIN REDBREAST. On the other hand, the old synonym RUDDOCK is too little known, even as ROBIN RUDDOCK, to be used without qualification. At the same time, not all regional or local names can as yet be entirely ignored, even in formal writing. In Advisory Leaflet 210, issued by the Ministry of Agriculture and Fisheries in 1934, we noted 'The Lapwing, Peewit or Green Plover', in number 211 of the same year 'The Barn or White Owl'. It could hardly be otherwise today, half a century on.

Polemics

We did not deem it profitable, in this work, to engage in polemics. Where our findings on a difficult name have already been published, the source is indicated, since space here does not permit an exhaustive discussion of such cases. For example, under GANNET we baldly state that this word formerly meant Goose and that its present meaning arose in connection with gannet fowling on Lundy Island, crucial matters which had escaped the attention of lexicographers and etymologists alike. But for the details of the philological analysis and the supporting documentary evidence, reference must needs be made to the primary study.

In other controversial, but less complex, cases, and these are in the majority, it is believed that the evidence adduced here will speak for itself. One notices, for instance, that etymologists have hitherto supposed the element *lag* in GREY LAG GOOSE to be somehow connected with the verb lag (behind), surely in ignorance of the fact that *lag* is an old farmyard call to, and a name for, a Goose; see J. Wright, *English Dialect Dictionary*, 1905. Only A. Newton, *Dictionary of Birds*, 1893–6—without having the advantage of being able to consult the *EDD*—got near the truth when he noticed 'the curious fact . . . that to this day flocks of tame geese in Lincolnshire are urged on by their drivers with the cry of *Lag 'em, Lag 'em'*. In such cases our interpretation is, in principle, given without reference to previous speculation, and the reader may assume that there are the best of reasons for rejecting contrary statements. We would add, too, that in instances where the record is demonstrably unreliable, we have not usually mentioned this in so many words. When we trace WOOD HEN back to Old English *wuduhenn*, we tacitly ignore the fact that, in the actual source, the word is glossed by Latin *coturnix* quail, surely an error, since by no stretch of the imagination could the Quail be called a woodland species. At the same time, where by scholarly standards justifiable doubt exists in any of these matters, the circumstances are stated as explicitly as may be; see e.g. GOOSANDER or MARTLET.

Sources

The following books have been our chief sources. Firstly, the *Oxford English Dictionary*, 1884–1933, and its *Supplement*, 1972–, with its rich documentation reaching back to the earliest period of the language. This material could sometimes be supplemented from new data available in the *Shorter Oxford English Dictionary* (ed. 3), 1973, or from the monumental *Middle English Dictionary*, 1954–, and occasionally also directly from older ornithological literature. Next, the *English Dialect Dictionary*, 1898–1905, a great repository of local names known to have been in use since 1700; it incorporates the collection found in Charles Swainson, *Provincial Names . . . of British Birds*, 1885. Scottish examples are covered in the extensive *Scottish National Dictionary*, 1931–76. Many more local English items can be culled from Christine Jackson, *British Names of Birds*, 1968, which, with variants, lists over 4,000 names.

The coverage provided by the above sources formed the foundation of the present work. Far-ranging though it is, it could not be called exhaustive, for doubtless many local names await a collector or publisher. This becomes apparent when one considers the names printed by John Braidwood in 'Local Bird Names in Ulster', *Ulster Folklife*, 1965, pp. 98–135, and in two supplements in the same journal: 1966, pp. 104–7, and 1978, pp. 83–7. Here is the most comprehensive body of data ever assembled from a comparable locality in these islands. The profusion of synonyms and the number of variants is astonishing. Many species go under half a dozen or more different names, any one of which may exist in numerous varying forms. Incredible though it may seem, the name Yellow Yorling, the usual Ulster expression for the Yellowhammer, occurs in some twenty-five variants—Jackson notes ten from the whole of the British Isles, Swainson five.

Not all our bird names are folk names, many having been created by naturalists during the past three hundred years or so. In this connection, H. Kirke Swann, *A Dictionary of . . . Names of British Birds*, 1913, often provides a welcome guide to details not found in any general dictionary. In the majority of these cases, however, the information required had to be sought for in the original works of the ornithologists.

As to the etymologies themselves, the primary sources are again the dictionaries referred to above. C. T. Onions, *Oxford Dictionary of English Etymology*, 1966, presents in summary form the findings of the *OED*, but does not go beyond these. Nor have other etymologists, such as Skeat (1909), Weekley (1920), or Partridge (1958), added anything of substance to the conclusions of the *OED* on the subject of bird names. These conclusions represent the achievements of the etymological science of the day and are, for the greater part, undoubtedly correct. A re-examination of the more problematic names, however, has shown that certain interpretations were in need of revision, while in cases where no etymology could formerly be given, a solution has sometimes been found. As regards the great mass of local words not figuring in the *OED*, etymological coverage is excellent as far as Scottish names are concerned, thanks to the *SND*. The purely English names, on the other hand, have not been so thoroughly treated, indeed quite a number became known too late for inclusion in the *EDD*. Nevertheless, that dictionary provides a reliable guide to many etymologies, which we have been able to supplement from our own researches.

It goes without saying that etymological studies of cognate names in the languages most closely related to English have often yielded valuable clues, in particular the standard monograph by H. Suolahti, *Die deutschen Vogelnamen*, 1909. By the same token, important additional information on names of French origin could be gleaned from the latest French dictionaries, both historical and etymological. In the case of names taken from Celtic, recent literature has likewise been consulted with profit.

BIBLIOGRAPHY

Albertus c.1250: Albertus Magnus, De Animalibus, c.1250.

Albin 1738: E. Albin, A Natural History of Birds, 1731–8.

Aldrovandi 1603: U. Aldrovandus (Aldrovandi), Ornithologia, 1599–1603.

Bechstein 1795: J. M. Bechstein, Gemeinnützige Naturgeschichte Deutschlands, 1789–95.

Belon 1555: F. Belon, L'Histoire de la Nature des Oyseaux, 1555.

Bewick 1797, 1804: T. Bewick, A History of British Birds, i, Land Birds 1797, ii, Water Birds 1804.

Bewick 1821: A Supplement to the History of British Birds, 1821.

Bewick 1826, 1832: Subsequent editions of Bewick 1797, etc.

BOU 1883: A List of British Birds compiled by a Committee of the British Ornithologists' Union, 1883.

Braidwood 1965: J. Braidwood, 'Local Names in Ulster', Ulster Folklife, 1965, pp. 98–135.

Brehm 1831: C. L. Brehm, Handbuch der Naturgeschichte aller Vögel Deutschlands, 1831.

Brisson 1760: M. J. Brisson, Ornithologia, 1760.

Browne 1646: T Browne, Pseudodoxia Epidemica, 1646.

Browne a.1682: Account of Birds . . . found in Norfolk, by T. Browne (died 1682), ms. ed. S. Wilkin, 1835.

Brünnich 1764: M. Th. Brünnich, Ornithologia borealis, 1764.

Buffon 1775: G. L. Leclerc, Comte, L'Histoire naturelle des Oyseaux, 1770–83.

Buffon 1792: Natural History (translation), 1792.

Caius 1570: J. Caius (Kay), De Rariorum Animalium . . . Historia, 1570.

Catesby 1731: M. Catesby, Natural History of Carolina, 1731.

Charleton 1668: W. Charleton, Onomasticon Zoicon, 1668.

Chaucer c.1381: G. Chaucer, Parlement of Foules, c.1381.

Checklist 1952: Checklist of Birds of Great Britain and Ireland, 1952.

Clusius 1605: C. Clusius, Exoticorum libri decem, 1605.

Cooper 1565: T. Cooper, Thesaurus, 1565.

Cotgrave 1611: R. Cotgrave, A Dictionarie of the French and English Tongues, 1611.

Dixon 1894: C. Dixon, The Nests and Eggs of British Birds, 1894

Drayton 1622: M. Drayton, Poly-Olbion, 1613–22.

Dryden a.1700: J. Dryden (d.1700), (translation of Chaucer's) Wife of Bath, a. 1700.

EDD: J. Wright, The English Dialect Dictionary, 1898–1905.

Edwards 1751, etc.: G. Edwards, Natural History of Uncommon Birds, 1743–51.

Edwards 1764, etc.: G. Edwards, Gleanings of Natural History, 1758–64.

Eyton 1836: H. Eyton, History of the Rarer British Birds, 1836.

Faber 1822: F. Faber, Prodromus der isländischen Ornithologie, 1822.

Fisher 1966: J. Fisher, The Shell Bird Book, 1966.

Fleming 1828, 1842: J. Fleming, A History of British Animals, 1828, ed. 2, 1842.

Florio 1598: J. Florio, Italian-English Dictionary, 1598.

Gaza 1476: Th. Gaza, *Aristotelis de Natura Animalium* (Latin translation of Aristotle), 1476.

Gesner 1555: C. Gesnerus (Gesner), *Historia Animalium Liber III qui est de Avium Natura*, 1555.

Giraldus *c*.1185: Giraldus Cambrensis, *Topographia Hibernica*, *c*.1185.

Gmelin 1793: C. Linnaeus, *Systema Naturae*, ed. 13, ed. J. F. Gmelin, 1788–93.

Gould 1834, etc.: J. Gould, *Birds of Europe*, 1832–7.

Halliwell 1847: J. O. Halliwell, *Dictionary of Archaic and Provincial Words*, 1847.

Hartert 1912: E. Hartert et al., *Hand-list of British Birds*, 1912.

Hett 1902: C. L. Hett, *Glossary of Popular, Local and Old-fashioned Names of British Birds*, 1902.

Hill 1752: J. Hill, *A General Natural History*, 1748–52.

Hodges 1643: R. Hodges, *A Special Help for Orthographie*, 1643.

Holland *c*.1450: R. Holland, *The Buke of the Howlat*, *c*.1450.

Holme 1688: R. Holme, *The Academy of Armory*, 1688.

Hutton 1956: H. P. W. Hutton, *The Ornithologists' Guide*, 1956.

Jackson 1968: Christine E. Jackson, *British Names of Birds*, 1968.

Jenyns 1835: L. Jenyns, *A Manual of British Vertebrate Animals*, 1835.

Johnson 1755: S. Johnson, *Dictionary of the English Language*, 1747–55.

Latham 1781, etc.: J. Latham, *General Synopsis of Birds*, 1781–5.

Latham 1787: *A Supplement to General Synopsis of Birds*, 1787.

Latham 1790: *Index Ornithologus*, 1790.

Latham 1802: *A Supplement to General Synopsis of Birds*, 1802.

Leach 1816: W. E. Leach, *Systematic Catalogue of . . . Birds . . . in the British Museum*, 1816.

Leslie 1578: J. Leslie, *Scotiae Descriptio*, 1578.

Linnaeus 1758: C. Linnaeus, *Systema Naturae* (ed. 10), 1758.

Lockwood 1961: W. B. Lockwood, *The Faroese Bird Names*, 1961.

Macaulay 1764: K. Macaulay, *Story of St. Kilda*, 1764.

MacGillivray 1840: W. MacGillivray, *History of British Birds*, 1837–52.

Martens 1694: F. Martens, *Voyage to Spitzbergen 1671*, 1694 (translated from the German).

Martin 1698: M. Martin, *A late Voyage to St Kilda*, 1698.

Martin 1703: *A Description of the Western Islands of Scotland*, 1703.

MED: H. Kurath and S. M. Kuhn, *Middle English Dictionary*, 1954–.

Merrett 1667: C. Merrett, *Pinax Rerum Naturalium Britannicarum*, 1667.

Minsheu 1627: J. Minsheu, *The Guide into Tongues*, 1627.

Montagu 1802: G. Montagu, *Ornithological Dictionary*, 1802.

Montagu 1813: *A Supplement to Ornithological Dictionary*, 1813.

Montagu 1831, 1866: eds. 2 and 3 of *Ornithological Dictionary and Supplement*, 1831, 1866.

Mouffet 1655: T. Mouffet, *Health's Improvement*, 1655.

Mudie 1834: R. Mudie, *British Birds*, 1834.

Nance 1963: R. M. Nance, *A Glossary of Cornish Sea-Words*, 1963.

Newton 1896: A. Newton, *Dictionary of Birds*, 1893–6.

OED: J. A. H. Murray et al., *A New English Dictionary on Historical Principles*, 1884–1933, later known as *The Oxford English Dictionary*.

OED Suppl.: R. W. Burchfield, *A Supplement to the Oxford English Dictionary*, 1972–.

Opie 1952: Iona and P. Opie, *The Oxford Dictionary of Nursery Rhymes*, ed. 2, 1952.

*Owl and N. a.*1250: *The Owl and the Nightingale*, *a.*1250.

Palsgrave 1530: J. Palsgrave, *Lesclarcissement de la Langue Francoyse*, 1530.

Palsgrave *c.*1532: (G. Duwes) *An Introductorie*, etc., *c.*1532.

Parry 1962: M. Parry, *Casgliad o Enwau Adar*, 1962.

Paston 1459: *The Paston Letters*, 1459.

Pennant 1768: T. Pennant, *British Zoology*, ii, 1768.

Pennant 1770: *Appendix to the British Zoology*, 1770.

Pennant 1773: *Genera of Birds*, 1773.

Pennant 1776: *British Zoology*, ed. 4, ii, 1776.

Pennant 1785: *Arctic Zoology*, ii, 1785.

Pennant 1812: *British Zoology*, ed. 5, ii, 1812.

Pliny *c.*77: Plinius (Pliny), *Naturalis Historia*, x, Aves, *c.*77.

*Prompt.Parv. c.*1440: *Promptorium Parvulorum sive Clericum, Lexicon Anglo-Latinum Princeps, c.*1440.

Ray 1661, 1662: J. Ray, *Itineraries* 1661, 1662, ms. ed. E. Lankester, 1848.

Ray 1674: *A Collection of English Words . . . with Catalogue of English Birds*, 1674.

Ray 1676: F. *Willughbeii . . . Ornithologia libri tres . . .* Totum opus recognovit . . . J. Raius, 1676.

Ray 1678: *The Ornithology of Francis Willughby . . .* Translated into English, and enlarged by . . . J. Ray, 1678.

Ray 1691: *A Collection of English Words*, ed. 2, 1691.

Ray 1713: J. Raius (Ray), *Synopsis methodica Avium et Piscium*, 1713.

Reaney 1976: P. H. Reaney, *A Dictionary of British Surnames*, ed. 2, 1976.

Scopoli 1769: G. A. Scopoli, *Annus (I) Historico-naturalis*, 1769.

Selby 1825, 1833: P. J. Selby, *Illustrations of British Ornithology*, i, Land Birds 1825, ii, Water Birds 1833.

Sibbald 1684: R. Sibbald, *Scotia Illustrata; sive Prodromus Historiae Naturalis*, 1684.

Skelton *c.*1508: J. Skelton, *Phylyp Sparrow, c.*1508.

Skinner 1671: S. Skinner, *Etymologicon Linguae anglicanae*, 1671.

SND: W. Grant and D. D. Murison, *The Scottish National Dictionary*, 1931–76.

Sowerby 1806: J. Sowerby, *British Miscellany*, 1804–6.

Stark 1828: J. Stark, *Elements of Natural History*, 1828.

Stephens 1817, etc.: G. Shaw, *General Zoology* (1800–12) continued by J. F. Stephens 1815–26.

Suolahti 1909: H. Suolahti, *Die deutschen Vogelnamen*, 1909.

Swainson 1885: C. Swainson, *Provincial Names and Folk Lore of British Birds*, 1885.

Swann 1913: H. K. Swann, *A Dictionary of English and Folk Names of British Birds*, 1913, repr. 1977.

Temminck 1820: C. J. Temminck, *Manuel d'Ornithologie*, 1820.

Tunstall 1771: M. Tunstall, *Ornithologia Britannica, seu, Avium omnium britannicarum . . . catalogus*, 1771.

Turner 1544: W. Turner, *Avium precipuarum, quarum apud Pliŋium et Aristotelem mentio est, brevis et succinta historia*, 1544.

Turner 1562: *A new Herball*, 1551–62.

Walcot 1789: J. Walcot, *Synopsis of British Birds*, 1789.

White 1767, etc.: G. White, *The Natural History and Antiquities of Selborne*, 1789—essentially the author's letters to Pennant and Barrington from 1767 and 1769 to 1780 and 1787 respectively.

Whitman 1899: C. H. Whitman, 'The Birds of Old English Literature', *The Journal of Germanic Philology*, ii, 149–98 (1898–9).

Wilkins 1668: J. Wilkins, *Essay* . . . (with an alphabetical dictionary), 1668.

Withycombe 1959: E. G. Withycombe, *The Oxford Dictionary of English Christian Names*, ed. 2, 1959.

Worm 1655: O. Wormius (Worm), *Museum Wormianum*, 1655.

Yarrell 1843, 1856: W. Yarrell, *History of British Birds*, 1843, ed. 3, 1856.

PERIODICAL PUBLICATIONS

Contributions by the author to the following:

Amster. St.: Amsterdam Studies in . . . *Linguistic Science*
BB: *British Birds*
BBCS: *Bulletin of the Board of Celtic Studies*
Fróð.: *Fróðskaparrit*
JMM: *Journal of the Manx Museum*
Neuphil.Mitt.: *Neuphilologische Mitteilungen*
SGS: *Scottish Gaelic Studies*
TPS: *Transactions of the Philological Society*
ZAA: *Zeitschrift für Anglistik und Amerikanistik*
Z.f.celt.Phil.: *Zeitschrift für celtische Philologie*

ABBREVIATIONS

a. ante, before
c. circa, about
cf. confer, compare
pl. plural
sg. singular

Beds.	Bedfordshire	Lincs.	Lincolnshire
Berks.	Berkshire	Norf.	Norfolk
Bucks.	Buckinghamshire	Northants.	Northamptonshire
Cambs.	Cambridgeshire	Northumb.	Northumberland
Ches.	Cheshire	Notts.	Nottinghamshire
Cumb.	Cumberland	Oxon.	Oxfordshire
Derbys.	Derbyshire	Shrops.	Shropshire
Glos.	Gloucestershire	Staffs.	Staffordshire
Hants.	Hampshire	Suff.	Suffolk
Heref.	Hereford	Warwicks.	Warwickshire
Herts.	Hertfordshire	Westmor.	Westmorland
Kirkcud.	Kirkcudbright	Wilts.	Wiltshire
Lancs.	Lancashire	Worcs.	Worcestershire
Leics.	Leicestershire	Yorks.	Yorkshire

NOTE TO THE READER

Where traditional names are found in older sources, we have added the date of each occurrence quoted. If the source is a treatise on ornithology or natural history, a well-known dictionary or work of literature, the name of the author (if known) and the work have also been given. When these details are not supplied, they may be sought in the *OED*.

A large number of names, especially the more local ones, were not recorded until recent times, many within the last hundred years. In most of these cases, there seemed little point in adding fortuitous dates, since all can be assumed to have been in existence by 1700 at the latest.

Certain local names have no fixed spelling, so that entries below are occasionally perforce arbitrary. In so far as non-standard names are concerned, we have generally preferred to write the parts separately, thus FURZE CHAT, but standard WHINCHAT.

Names printed in small capitals refer to other entries.

A

Aberdevine. An obsolete term for the Siskin, reported from Ches. and the Beverley district of Yorks., said to be a bird-catcher's name. It is first mentioned by Albin 1738 'Abadavine, Aberduvine, Aberdevine'. Pennant 1768 knew 'Aberdavine' as a name heard in London bird shops. No clue as to the possible origin of this peculiar expression has been found so far.

It is probably an artificial word, because the Siskin is mainly a bird of the north of Scotland and of Ireland, where until modern times the usual language was Gaelic. In these circumstances, it is unlikely that the wild bird ever acquired a genuine English folk-name. If noticed at all, it will presumably have been regarded as a small Greenfinch, in which connection see BARLEY BIRD.

Aichee, see HAY SUCK.

Alamootie, Alamottie. Variant names for the Stormy Petrel from Shetland and Orkney respectively, the former reported as early as 1774, though miswritten 'Alamonti'. These words are basically Norn. The first element *ala-* means fat. As to the second, the more original variant is Shetland -MOOTIE consisting of **moot-* little thing, with the redundant diminutive suffix -IE. A name of this sort would, in the first place, denote the corpulent nestling, formerly much sought after for its body oil; the same motivation is found in MARTIN OIL.
ZfAA xix 57f. (1971)

Alan (Allan). A widespread Scottish name for the Arctic Skua, also used in Orkney and Shetland. The record goes back to 1806, but the name occurs earlier in the commonly heard alternatives DIRTEN ALAN, SCOUTIE ALLAN. The word is felt to be the Christian name, so long a favourite in Scotland, but this will hardly be the true origin, which remains unknown.

Alf, see ALP.

Allan, Allan Hawk, see ALAN, ARRAN HAWK.

Alp. The oldest known name for the Bullfinch, the first attestation 'albus' pl. being datable to 1352–53, then *a.*1400 'alpes', next *Prompt.Parv. c.*1440 'Alpe'. We further note Ray 1674 and 1678 'Bulfinch, Alp or Nope', the last two described as local or regional. In recent times, the word is known from East Anglian and north country sources, the former having a variant **Olp.** For an unknown reason, East Anglian forms may change *p* into *f*, hence also **Alf, Olf,** further with loss of *l*: **Awf.**

The history of this name is beset with obscurities. Nothing is known of its origin. The first attestation with *alb-* as opposed to the somewhat later **Alp** adds to the uncertainty about the exact shape of the original; in fact, both forms are phonologically unusual in the English of the day—and the same applies to the contemporary French, so that there is no special reason to regard the word as ultimately from that quarter. Nor do later forms appear to help. The type *alb-* emerges again in an isolated record of 1576 as **Awbe,** i.e. with philologically plausible loss of *l*, cf. **Awf** above, further **Ope** (see NOPE).

Anchor Bird. A Sussex term for the Swift, inspired by the silhouette of the bird gliding aloft.

Annet. An apparently now disused name, reported as meaning Kittiwake in Yorks., but Common Gull in Northumb. It will have been a pet name, simply the old-fashioned Christian name, once so common in the North, and also seen in PIANNET.

Apple Bird. A name in Cornwall for the Chaffinch, deriving by popular etymology from SHELD APPLE or one of its modifications.

Arctic Skua. Coined by Fleming 1828, the epithet doubtless suggested by earlier naturalists' terms: Pennant 1776 'Arctic Gull', Edwards 1750 'Arctic Bird'. It for long competed with RICHARDSON'S SKUA, but, chosen by Hartert 1912, it came to prevail as the standard name a few years after that. This species has several folk names, as ALAN, DUNG BIRD, SHOOIE, TEASER.

Arctic Tern. Temminck 1820 separated this species from the Common Tern, naming it *Sterna arctica*, after the analogy of which Fleming 1828 introduced the present English term.

(-ard. A suffix of French provenance, arriving in Norman times in such names as BUZZARD or MALLARD. It sometimes changed locally to -*art*, both forms being regularly weakened in pronunciation (hence sometimes written -*erd*, -*ert*) when interchange with names ending in -ET became easy, cf. LINNARD, LINNART from LINNET. The suffix occasionally changes with -ER, notably in POCHARD beside original POKER/ POACHER, and can consequently acquire a name-forming function, as in WINNARD.)

Arling. Given as a name for the Wheatear in Turner 1544 '*caeruleo* a clotburd, a smatche, an arlyng, a steinchek' and occasionally quoted by later writers, the last instance dating from 1753, but these are dependent on Turner. Apparently to be analysed *ar-* plus -LING, the latter being the well-known suffix, but the first part seems intractable. Turner's book was produced in Cologne and contains numerous misprints, of which 'clotburd, steinchek' above are certain examples, cf. CLOT BIRD, STONE CHACK. It is possible that the present name, which bears no resemblance to any other Wheatear name, is in fact spurious.

Arran Hawk. A term for the Great Northern Diver known from Ayrshire and Argyll and from the Irish coast opposite. It is an Anglicization of a lost

Gaelic **aranag*, the diminutive of **arana*, an echoic formation comparable to synonymous CARARA. The present form appears further corrupted in ALLAN HAWK, i.e. through association with the local Skua name ALLAN, whence by another corruption HOLLAND HAWK.

Arse Foot. A Grebe name, first in Florio 1598 '*Giuero* . . . a bird called a diuer, didapper, or arsefoot' and used by some early naturalists, e.g. Merrett 1667 '*Columbus medius* Dive-dapper, or Arsfoot'. Hardly a folk-name, but a book word suggested by Gesner 1555 'Mergulus . . . arsevoet', i.e. now obsolete Dutch *aarsvoet*. In the 16th cent., the first element of this name was not yet felt to be vulgar, in which connection compare WHEATEAR.

(-art, see -ARD.)

Ash-coloured Falcon. Montagu's name for the species identified by him in 1802 and previously not distinguished from the Hen Harrier, now known as MONTAGU'S HARRIER.

Ash-coloured Harrier. A name for Montagu's Harrier coined by Selby 1825, a modification of ASH-COLOURED FALCON for systematic reasons.

Assilag. A Hebridean name for the Stormy Petrel, first in Martin 1698 and from this source often quoted in the literature. It is an approximate spelling of Gaelic *asaileag*, of uncertain origin beyond that it contains the common diminutive suffix -*ag*.
  SGS xii 277 (1976)

Atteal. An obsolete name from Orkney and Shetland denoting some kind of small Duck wintering in the islands. Attestations, often associated with TEAL, start in 1599 'Teillis, Atteilis' and include 1652 'teales and awteals', 1653 'Taile or Ataile', 1809 'A-teal', 1813 'Atteal'. Antecedents are obscure. One can only note a superficial resemblance to Danish *atling* garganey, a word most likely having as its basic sense small duck, and imagine that a Norn word of this shape was once in use in Orkney and Shetland. In modern times, this hypothetical Norn term was then presumably modified by contact with English TEAL.

Auercalze. Sometimes quoted as an old name for the Capercaillie. It is, however, spurious, being in fact a misprint; see under CAPERCAILLIE.

Auk. First recorded as a northern word for the Razorbill in Ray 1674 'Rasor-bill: Auk, or Murre', terms repeated in Ray 1678. Pennant 1768 adopted the present name in this traditional sense, but at this same time, inspired by Linnaeus 1758 *Alca*, also used it generically. In 1770, he removed the inconsistency by substituting 'Razor-bill', an arrangement acceptable to his successors, hence our present-day usage. Except in the proverb 'drunk as an auk' (Swainson 1885), this word has in the meantime died out as a folk-name, doubtless being confused with HAWK, which in ordinary speech often loses its aitch.

Our word is a relic of the language of the Vikings, continuing Old Norse *alka*, the original meaning of which is neck. It is thus a part for whole name and alludes to the Razorbill's peculiar habit of stretching its neck after returning to the rocks. This stretching is accompanied by swaying movements of the body which motivated the proverb. See FALK.

Fróð. xix 126–9 (1971), *Neuphil. Mitt.* lxxix 391–5 (1978)

Avocet. A term introduced into ornithological literature by Aldrovandi 1603 as *avosetta*, a name from the Venetian coast, the Italian habitat of this species. Its etymology is unknown. It was employed as the scientific name in Charleton 1668, also Ray 1678 'Avosetta of the Italians', and adopted by Linnaeus 1758 *Recurvirostra avosetta*. Pennant 1768 coined the expression 'scooping avoset', but in 1776 he put simply 'avoset', thus *de facto* establishing the exotic name as standard English. The present spelling dates from Selby 1833 under the influence of French *avocette*, due to Brisson 1760 who apparently misspelt the original.

We presume that Pennant preferred this foreign word since the genuinely English names of this bird are rather inelegant. Pennant himself quotes CROOKED BILL and SCOOPER; other synonyms include BARKER, CLINKER, COBBLER'S AWL, YELPER, none exactly decorous.

Awbe, Awf, see ALP.

Awl Bird, see HOOD AWL.

B

Baakie. A name for the Greater Black-backed Gull from Orkney and Shetland. A Norn term, it continues the Old Norse root *bak-*, the ending -IE being a relatively recent addition. The Old Norse source *bakr* (best surviving today in Faroese *bakur*) is an aphetic form of *svartbakr* literally black back, evidently an innovation of the Orkney-Shetland-Faroe area. Cf. SWAABIE.

Bag. A Northants. name for the Long-tailed Tit, properly a word referring to the bag-shaped nest, cf. POKE BAG, LONG POD.

Bald Buzzard. Used of the Marsh Harrier and the Osprey, but properly a name for the Kite, as in Merrett 1667 'Peronos bald Buzzard, or Kite'. In popular use BUZZARD and KITE are often interchangeable, as explained under BALD KITE.

Bald Coot. A traditional name for the Coot, in popular use in Sussex, Somerset and Glos., known since *a*.1325 'Une blarye, a ballad cote'. The epithet *bald* naturally alludes to the featherless white shield. For a northern counterpart, see BELL KITE. There is comparable motivation in SNYTH.

Bald Kite. The name KITE has been popularly used for both the Kite and the Buzzard, but the former could be distinguished by the present term, *bald* referring to the white head of the Kite. Even so, names and species, as so often, continued to be confused, as already shown in the first known occurrence in Cotgrave 1611 'Buzzard, or Bald-kite'.

Bank Martin. A name for the Sand Martin, first in Charleton 1668 'Sand, or Bank Marten', Ray 1678 (Index) 'Bank Martin'; it was used by White 1774 'sand-martin, or bank-martin'. This species often chooses suitably steep river-banks into which to drive its nesting tunnels. The name has almost certainly

been modelled on BANK SWALLOW. It survives today in Wilts.

Bank Swallow. A name for the Sand Martin, the records going back to Mouffet 1655. *Bank* here means specifically river bank, the commonest site for the birds' nesting burrows. Occurring nowadays in Worcs. and Yorks., the term was doubtless once widely used and has, in fact, become the usual name for this species in American English. It is, in essence at least, the oldest surviving English name for this bird: it reflects Old English *stæpswealwe* literally bank swallow, first found in the 10th cent., *stæp* being replaced during the Middle English period by *bank*, a word of Norse origin. Cf. BANK MARTIN, and see also MARTIN.

Bar Drake, Bar Duck, Bar Gander, see BERGANDER.

Bark Creeper. A name from an unspecified locality for the Tree Creeper, also similarly motivated Suff. **Bark Runner** and Scottish **Bark Speeler** (climber), cf. TREE SPEELER.

Barker. A Suff. name for the Black-tailed Godwit, inspired by the loud, insistent voice. Also a local name for the Avocet, suggested by the alarm call. Cf. YELPER.

Bark Runner, Bark Speeler, see BARK CREEPER.

Barley Bird. A term applied very locally, from some real or fancied association with the barley crop, to such diverse species as the Greenfinch, the Common Gull, the Nightingale, the Grey and Yellow Wagtails, and the Wryneck. Ray 1678 reports it as a Sussex term for the Siskin, apparently confused with the Greenfinch.

Barnacle. The name in this form is first noticed in 1598, having developed by regular change from **Bernacle** (for the phonetics, see BERGANDER) known since the first half of the 15th cent. This earlier form was kept by Ray 1678 and Pennant 1768, to be often copied by later authorities, but the present form became usual during the 19th cent. The name is paralleled in French *barnacle*, *bernacle*, and contemporarily in Latin *bernacula*. A variant **Bernicle**, in occasional use until the beginning of this century, goes back to Turner 1544, cf. French *bernicle* and the corresponding Latin *bernicla*.

The name exists in a still older form in early 13th cent. 'bernekke', early 14th 'bernak', continuing into the 15th. These too have counterparts in French *bernaque* and Latin *bernaca*, in which shape the word first comes into our ken in Giraldus *c.*1185. Giraldus recounts the legend that this bird grows out of a sort of shell, a legend which, in various forms, was believed even as late as the 17th cent. No adequate explanation of the superstition has yet been given.

It is apparent that the English and French languages have here been in contact, and further that the vernacular forms are simply book words based upon Latin *bernacula*, now seen to be a relatively recent diminutive of *bernaca*. Though conventionally taken as meaning Barnacle Goose, this name must in practice have included the Brent, since the two species were often confused, as Ray 1678 noted; *bernaca* has therefore the general sense of Black Goose. A review of Black Goose names in different languages shows that these are nearly always imitative of the cry or else allude to the dark plumage. But *bernaca* does not appear to fit into either of these categories. In view of the legend, however, one may consider the possibility that the name primarily denotes the shell. Giraldus uses the term in connection with Ireland, where *bairneach* is the Gaelic term for a limpet, which could well have been Latinized as *bernaca*. Admittedly, a limpet is not a barnacle, but given the credulous milieu, this detail need not detract.

For another version of the legend, see TREE GOOSE.

Barnacle Goose. Forms of the original name BARNACLE qualified by GOOSE have tended to be regarded as standard since Pennant 1768 introduced **Bernacle Goose**, the simplex being of course ambiguous, as it can also denote a cirriped.

Barn Owl. So called from a predilection for nesting in barns; it has been documented since Ray 1678 'The common Barn-Owl, or White-Owl, or Church-Owl'. From these, Pennant 1768 selected 'White Owl', but the present name was often used by his successors, sometimes as an alternative to Pennant's choice. Yarrell 1843 and later editions still admitted both, but in his text he shows a clear preference for Barn Owl, officially adopted in BOU 1883.

Barred Woodpecker. A term for the Lesser Spotted Woodpecker coined by Bewick 1797. The black and white plumage on the back of this species presents a barred appearance, in this respect distinguishing it from the Great Spotted which has white bars on the wings only. Bewick's term has sometimes found favour, especially in this century; see SPOTTED WOODPECKER.

Barrel Tit. Given by Swann 1913 as a provincial name for the Long-tailed Tit. Barrel refers to the shape of the astonishing nest; cf. BUM BARREL, also BOTTLE TIT.

Bar-tailed Godwit. Coined by Fleming 1828 and taken up by subsequent writers. The species had previously been called **Common Godwit**, and before that simply GODWIT, as Pennant 1768 following Ray 1678 'Godwit . . . Yarwhelp, or Yarwip'; see also BLACK-TAILED GODWIT.

Bass Goose. A local Scottish term for the Gannet in reference, of course, to the birds stationed on Bass Rock in the Firth of Forth. Although not noticed until

Swainson 1885, this name must be much older, witness Turner 1544 whose Latin *Bassani anseres* surely translates Bass Geese. The literary Scots name is SOLAN GOOSE.

Basser. A Forfar name for the Gannet from Swainson 1885; see BASS GOOSE, also -ER.

Beak. Introduced by the Normans, this word is recorded in an English text as early as *a*.1250 'bec'. It appears *c*.1400 as the name of a bird, probably the Woodcock or Snipe: '*Et le oisel qi ad noun bibakaz* . . . And the byrd that hath name beke' (*MED*). In this function, it was most likely a fowler's term; see further under WOODCOCK. On part for whole names, see SHAG. The source is Old French *bec* from Late Latin *beccus* of Gaulish provenance. Beak is today the standard expression in ornithological work, but in other spheres it competes, sometimes successfully, with BILL. Nevertheless, it must for long have been something of a bookish word, for unlike the native terms BILL and NEB it does not enter into the formation of bird names, except as above and in the special case of GROSBEAK, and an *ad hoc* creation CROSSBEAK. See BECK.

Beam Bird. A name used in many parts of the country, from Northumb. to Wilts., for the Spotted Flycatcher, which often chooses a beam in an outhouse against which to build its nest. Cf. WALL BIRD.

Bean Goose. Originally a Lincs. name, first occurring in Pennant 1768, with the explanation that it was motivated by the likeness of the nail of the bill to a horse bean. This bird is commonly found in cultivated fields, foraging for grain, pulses, or potatoes. It is thus more likely that the present name derives from a feeding habit, especially in the light of the synonym CORN GOOSE. See PINK-FOOTED GOOSE.

Bearded Tit. Generalized by Yarrell 1843 for the older **Bearded Titmouse**, used by Pennant 1776 and apparently coined by Tunstall 1771. The name is, in fact, a translation of Brisson 1760 '*Parus barbatus*', itself inspired by French *mésange barbue*, this species being at the time classified as a member of Tit (*Parus*) tribe. The adjective bearded refers to the black facial markings of the male bird. For a traditional name, see REED-LING.

Beck. An East Anglian term for the Shoveler, a part for whole name, being an archaic form of BEAK.

Bee Eater. A summer vagrant without a genuine English folk-name. The present term, first in Charleton 1668, is a translation of ornithological Latin *apiaster* (*apis* bee).

Bell Kite. A northern, especially Scottish, form of BALD COOT, first *c*.1450 'Beld cyttes'. As to etymology, *bell*, originally *beld*, shows the regular regional development of *bald* (cf. local *peth* for *path*), but the second half of the modern form is corrupt, due to contamination from the unrelated name KITE. Cf. QUEET.

Bell Poot, Scottish **Bell Pout.** A Coot name in Northumb. and Linlithgow. For *bell*, see BELL KITE. The element *poot*, *pout* shows the local development of archaic English *poult* chicken, known since the 14th cent. It is a borrowing of Old French *poulet*, a diminutive of *poule* hen, from Latin *pulla*.

Bergander. A name for the Shelduck, reported from places as far apart as Northumb., East Anglia, and the Isle of Wight. The first record appears, in the present spelling, in Turner 1544, the next in 1570 'Bargander'. This latter shows the normal evolution of the former in the spoken language (just as *clerk* is nowadays pronounced /clark/, and in fact so written in the surname Clark); the beginnings of this phonetic change date back to the 14th cent. Needless to say, the new form was at once interpreted as **Bar Gander**. It follows, of course, that the many attempts to find an explanation of *bar*, e.g. as a bar of sand or a harbour bar, are in the nature of the case entirely vain.

In spite of its size, the species in question belongs to the Ducks, and is commonly named accordingly. As a consequence, **Bar Gander** was in some places transformed into **Bar Drake**, which in turn often further changed to **Bar Duck** in line with other Duck names. It is now evident that SHEL-DRAKE and its more recent synonym SHELDUCK can be seen as products of similar analogies, again with the exceptional use of DRAKE as the primary element in a name for a species of Duck. Since SHELDRAKE was in existence as early as 1195, its model Bergander must be older still, i.e. centuries older than the first record of 1544.

As this is an ancient name, we are more ready to accept -*gander* at its face value, in which case *ber*- is most likely to be an epithet appropriate to the male as opposed to the female. Practically the only visible feature distinguishing the sexes is the vivid, berry-red horn on the beak of the male. We therefore interpret *ber*- as a continuation of Old Norse *ber* berry, and observe that the very word as well as the thing reminds one of the berry, often so called, of the Mute Swan. At the same time it becomes clear why GANDER, and not GOOSE, came to be used as the primary name for the species.

Bernacle, Bernicle, see BARNACLE.

Bessy Blakeling, see BLAKELING.

Bessy Doucker, Bessy Ducker. A local name for the Dipper in the north and in Scotland. On the differing spellings, see DUCKER.

Bewick's Swan. This swan, at one time not distinguished from the Whooper, was shown to be a separate species by Yarrell in 1830. He proposed the name *Cygnus Bewickii*, i.e. Bewick's Swan, in memory of Thomas Bewick, the wood engraver, especially celebrated for his *History of British Birds*, who had died in

1828. Yarrell's proposal gained immediate acceptance.

Big Throstle. A northern name for the Mistle Thrush, alluding to its size; it is the largest of our Thrushes. See THROS-TLE.

Bilcock. A Yorks. name for the Water Rail, first in Ray 1678 'The Water-Rail, called by some the Bilcock or Brook-Ouzel'. In more recent times, the name has been found elsewhere in the north country with the meaning Moorhen, in which sense a variant **Bilter**, known since Skinner 1671, is also met with. This latter suggests a root *bilt*- extended by the agent ending -ER. That being so, Bilcock would originally be **Biltcock*.

The root in question will be an unrecorded verb **bilt*, a thinned form of **bult* from Middle English *bulten* to move quickly, bolt. The names thus primarily denote the Water Rail, being motivated by this bird's habit of swiftly running from one piece of cover to the next, as also in the case of other Water Rail names; see RUNNER.

(Bill. A word going back to medieval times: Middle and Old English *bile*, known since *c*.1000. It is not found in the other Germanic languages, but it appears to be cognate with Old English *bill* chopper (as still in bill-hook, hedging bill), a word with congeners on the Continent, as Dutch *bijl*, German *Beil* axe. The present word is apparently a more recent term than NEB, which it largely replaced. At the same time, it has had to compete, not always successfully, with BEAK; it cannot, for instance, be used of the beak of a bird of prey. Bill very often occurs as the second element in bird names, as RAZORBILL, SPOON-BILL.)

Billy Biter. A Sussex name for the Blue Tit. The sitting bird will peck at the fingers of bird-nesting boys. Cf. BITING TOM.

Billy Whit, Billy Wix. Norf. names for the Barn Owl, basically onomatopoeic, reproducing the sharp cry.

Bilter, see BILCOCK.

(**Bird.** A word of unknown affinities, first found *c*.800 as *brid*, thence by metathesis *bird*, the earliest occurrence being *c*.1000, both forms continuing side by side into the 16th cent., with *brid* predominating until *c*.1475, and remaining in some northern dialects. In Old English, the meaning was always nestling, young bird. This primary sense remained common until the end of the Middle Ages and survives to the present day in northern speech in the phrase 'a hen and her birds', and in swan-keeper's parlance, where bird means a cygnet. From about 1300, the word came to be applied to any small bird (see BLACKBIRD), such usage continuing into the 18th cent., when Johnson 1755 noted 'in common talk, *fowl* for the larger, *bird* for the smaller kind of feathered animals'. See FOWL.)

Bishop. A Scottish name for the Great Northern Diver, first in Martin 1703. The names of this species have often been tabooed, and it seems likely that the present facetious nickname arose in that connection. Martin's informants were Gaelic-speaking, so that Bishop is really a translation of Gaelic *easbuig*.

Bistard, see BUSTARD.

Biting Tom. A Blue Tit name from Northumb.; for motivation, see BILLY BITER.

Bitter Bank. A Borders name for the Sand Martin. The word *bitter* is to all appearances a development from Middle English *bīt the* bite the, assimilated to *bīt te*, subsequently taken as one word, when the vowel would be regularly shortened giving **bitte*, automatically equated with *bitter*. The bird excavates a nesting tunnel in a sandy bank, making this picturesque Bite-the-Bank an acceptable interpretation. The name was sometimes further changed to **Bitterie**, i.e. adding the diminutive suffix -IE, whence by folk etymology also a further development to **Butterie**.

Bitter Bump, see BUMBLE.

Bitterie, see BITTER BANK.

Bittern. A name borrowed from Old French *butor*, first in English *c*.1330 'botors'. Subsequently, the vowel often appears as *i*, thus Chaucer (*Wife of Bath*) *c*.1386 'bitore', Skelton *c*.1508 'The bitter with his bumpe'; Turner 1544 has 'bottour' and (Latinized) 'buttourum, bittourum', Browne 1646 'Bittor'. From these come the modern **Bitter Bump**, **Butter Bump**.

Since the early 16th cent., the type represented by 'Bittor' (above) also appears with final *n*, first 1515 'bittorn', then Bible of 1611 'bitterne', Ray 1678 'Bittour or Bittern or Mire-drum'. Pennant 1768 chose the present form, which has since been treated as standard.

The source word, French *butor*, has been known since the 12th cent. The Latin name was *būteō*, cf. the verb *būtīre* to boom like a bittern, based on the onomatopoeic root *būt-* reproducing the resonant call. This booming also earned the bird the Latin nickname *taurus* literally bull (cf. BULL O' THE BOG). It seems that French *butor* arose as a conflation of *būt-* and *taurus* (Old French *tor*).

Black Back. A name, easily created locally, to denote a Black-backed Gull. Cf. SADDLE BACK.

Blackbird. First mentioned in 1486, this name cannot be older than *c*.1300, since it reflects that stage in the evolution of the meaning of the word BIRD when it could apply to small birds, but not yet to large ones. When this name was created, there was thus no possibility of confusion with such birds as Crows or Ravens, which at the time would have been referred to as 'black fowls'. This neologism effectively replaced the traditional name OUZEL in the standard language during the course of the 17th cent. The same motivation is seen in COLLY BIRD.

Blackcap. A widespread popular name applicable to quite different birds. These include, in addition to *Sylvia atricapilla*, the Reed Bunting (see also SEAVE CAP) and the Black-headed Gull, going back to Ray 1678 'Pewit or Black-cap' (see PEWIT), further several species of *Parus*, namely Coal Tit, Great Tit, and Marsh (*cum* Willow) Tit, in which last sense it was recorded by Charleton 1668, also Ray 1678 'Marsh-Titmouse or Black-cap'. Pennant 1768 was the first author to use the name in its now standard meaning and his successors followed this example. Many naturalists, however, enlarged the name as **Blackcap Warbler** in their headings and tables, but this was never more than a nod in the direction of systematics, and has become less usual in this century.

Black Curlew, see GLOSSY IBIS.

Black Diver, Black Duck. Self-explanatory Scoter names, the former from Northumb., first noticed in Ray 1678 'Black Diver, or Scoter', the latter East Anglian and doubtless an old name too.

Black Game. A name for the Black Grouse, making its debut in Ray 1674, then 1678 'Heathcock or Black game or Grous'. Cf. RED GAME.

Black Goose. An Essex name for the Brent Goose. At the same time, it is commonly used as a general term embracing both the Brent and the Barnacle; ornithologists would also include the Canada Goose.

Black Grouse. A name going back to Ray 1674, then 1678 'Heathcock or Black game or Grous'. Of these three names, the present was that most generally used by the early naturalists, with Yarrell 1843 'Black Grouse' setting the seal of final approval.

Black Guillemot. Coined by Pennant 1768 in view of the bird's affinity with the Common Guillemot (Pennant simply 'Guillemot'), the inspiration coming from Brisson 1760 '*Uria* guillemot, *Uria minor nigra* petit guillemot noir'. The name was, of course, prompted by the summer plumage. It was used by nearly all succeeding authorities. Folk-names include DOVEKIE, TYSTIE.

Black Head, Black-headed Gull. Pennant 1776 called this bird Black Head, Latham 1785 expanding it to Black-headed Gull, from then on regarded as the standard term. For folk-names, see BLACKCAP, PEWIT.

Black Ibis, see GLOSSY IBIS.

Blackie. Widely used pet form of BLACK-BIRD; see -IE.

Black Martin. A name for the Swift, first recorded in Ray 1678 and, like synonymous **Black** (in Scotland **Brown**) **Swallow**, widespread in local use. Such terms were clearly inspired by the Swift's sombre plumage, which contrasts with the brighter colouring of the otherwise not so dissimilar Martins and the Swallow. Cf. BLACK SCREECH, COLLIER.

Black Neb. A Northumb. name for the Carrion Crow; see NEB.

Black-necked Grebe. This name appears in Hartert 1912 replacing EARED GREBE in use since Pennant 1768. It was a belated vernacular adjustment to the scientific *Podiceps nigricollis* (*nigri-* black, *-collis* necked) of Brehm 1831.

Black Ouzel. A name for the Blackbird, widespread in Lancs. and Yorks., and also found in Worcs. It is first noticed in Turner 1544 'blak osel, blak byrd'. The epithet *black* is a secondary addition, since OUZEL by itself traditionally denotes the species in question.

Black Puffin, see PUFFINET.

Black Redstart. Until this century no more than an occasional visitor and as such without a traditional name. The present term was coined by Gould 1834, who had identified the bird as a British species in 1829; it won immediate acceptance.

Black Scoter. The name given by Fleming 1828 to the Common Scoter and used by other naturalists until Yarrell 1843 introduced the term current today.

Black Screech. A Somerset name for the Swift combining the motives of colour and call; see BLACK MARTIN, SCREECH MARTIN.

Black Starling. An East Lothian term for the adult Starling, the glossy black plumage of which contrasts with the greyish-brown of the immature bird, in this locality known as the **Grey Starling**, elsewhere (unspecified) as the **Brown Starling**.

Black Swallow, see BLACK MARTIN.

Black-tailed Godwit. This species was first distinguished by Ray 1678 as 'the second sort of Godwit'. Edwards 1750 introduced **Red-breasted Godwit**, abbreviated by Pennant 1776 to **Red Godwit**, and this remained usual until replaced by the present name from Fleming 1828.

Black Tern. Coined by Pennant 1768, evidently with an eye to systematics, becoming the accepted term. For a traditional name, see STARN.

Black-throated Diver. Coined by Pennant 1776 as a counterpart to his RED-THROATED DIVER, and used by subsequent writers. It is doubtful if this species was ever prominent enough in our latitudes to acquire an English folk-name of its own.

Blakeling. A north country name for the Yellowhammer, *blake-* (from Old Norse *bleikr*) being a local word for yellow and *-ling* the name-forming suffix; in Yorks. and Westmor. also **Bessy Blakeling**. For a comparable formation, see YOULRING.

Blind Dunnock. A Somerset name for the Hedge Sparrow, apparently

because it cannot see that the Cuckoo lays an egg in its nest.

Blood Hoop. A Somerset name for the Bullfinch, comparable to East Anglian **Blood Olf** or **Blood Olp**, the epithet alluding, of course, to the crimson-breasted male. See further under HOOP 1, ALP.

Blue Back. A name for the Fieldfare, suggested by the blue-grey on the lower part of the back; it is in local use from Lancs. to Shrops., with an early attestation in Palsgrave 1532 'blewe back'. Cf. GREY THRUSH.

Blue Bonnet. A local name in Scotland for the Blue Tit, in fact the Scottish counterpart of BLUE CAP.

Blue Cap. A local, especially northern name for the Blue Tit. The term primarily denotes a cap of blue material, but it was also applied to a servant or tradesman, who traditionally wore such a cap. It is first mentioned in the second half of the 16th cent., going out of fashion by 1700. The bird name must therefore be at least as old as the latter date, though it is not actually recorded until 1804.

Blue Dove. A Yorks. (North Riding) term for the Rock Dove. The epithet *blue* is hardly specific, since it could equally well apply to the Wood Pigeon or the Stock Dove. The name might be envisaged as contrasting with TURTLE DOVE, but this species is hardly an everyday occurrence in the North Riding. It is just possible that the present term is ultimately of Norse origin, since comparable forms occur in some Scandinavian languages, as Faroese *bládúgva* rock dove, where *blá-* is redundant.

Blue Dunnock. A Ches. name for the Hedge Sparrow, *blue* having been added to DUNNOCK literally dun bird, after the analogy of other Hedge Sparrow names, as **Blue Isaac**, BLUE JIG; see BLUE SPARROW.

Blue Felt, also **Grey Felt.** Names for the Fieldfare, the former from Surrey, the

latter from Notts., the adjectives referring naturally to the indeterminate
blue-grey plumage; cf. PIGEON FELT and
see further under FIELDFARE.

Blue Hawk. A term applied locally to
various species, as the Hobby, Merlin,
Peregrine, Sparrowhawk; also the male
of the Hen Harrier.

Blue Isaac, see BLUE DUNNOCK, BLUE
SPARROW, HAY SUCK.

Blue Jig. A name for the Hedge Sparrow,
locality unspecified, with *blue* as in
other Hedge Sparrow names; see under
BLUE SPARROW. The word *Jig* is properly
Jug, occurring in synonymous HEDGE
JUG; details under JUGGY WREN.

Blue Neb. A Northumb. term applied to
both the Scaup and the Tufted Duck;
see NEB.

Blue Sparrow. A Scottish name for the
Hedge Sparrow. *Blue* is a familiar
epithet in Hedge Sparrow names, motivated by the bluish underparts, cf. **Blue
Isaac,** BLUE JIG, even BLUE DUNNOCK.

Blue Tit. Generalized by Yarrell 1843, for
the older **Blue Titmouse,** used by
Pennant 1768 and first found in Ray
1678 'The blue Titmouse, or Nun'. The
adjective *blue* also occurs in various local
names for this bird, as BLUE BONNET,
BLUE CAP, but unlike these the present
term is not indigenous. It is a translation
of Gesner 1555 *Parus caeruleus*, introduced by Ray for systematic reasons.
Gesner's Latin, however, is based on
the common German name *Blaumeise*.

Bob, Bobby. Local names for the Robin
from Notts. and Somerset, simply nicknames formed from ROBIN.

Bobby Wren. An East Anglian term for
the Wren, motivated by the bird's
distinctive bob-tail, cf. CUTTY WREN,
etc., the use of (a pet form of) a
Christian name as a prefix being, of
course, quite common, as in JENNY
WREN, etc.

Bob Robin. A tautological name for the

Robin, known from various parts of
England and Scotland; see BOB.

Bobwhite or **Bobwhite Quail.** A
North American game-bird introduced
in 1813 and on a score of occasions
since. Although it has bred here in the
wild, attempts at naturalization have
not apparently been permanent so far.
Commonly characterized as a Quail, the
bird takes its specific name from its
note, as one learns from the first
mention we have in 1812 'The quail will
sometimes sit, repeating, at short inter-
vals "Bob White", for half an hour at a
time' (*OED Suppl.*).

Bog Blutter, Bog Bumper, see BUMBLE.

Bog Drum. A local Scottish expression
for the Bittern, patterned on MIRE
DRUM, the replacement *bog* being common to other recent Bittern names; see
BUMBLE.

Bohemian Chatterer, see WAXWING.

Bonxie. This term, often used by writers
to denote the Great Skua, is proper to
the Shetland Islands. It was first recorded in 1774, but would have been
more correctly spelt **Bunxie*. The word
is Norn *bunksi*, formed from *bunke*
dumpy body, with a productive suffix
-si; cf. SPENSIE. The name thus literally
means dumpy bird; it is a neologism
which has replaced an older local term
Skooie (see SKUA) which would come
under pressure from the similar-
sounding SHOOIE Arctic Skua.

The present name may naturally be
preferred to GREAT SKUA by those who
would separate this species generically
from the other Skuas.

Boomer, see BUMBLE.

Boondie. An Orkney name for the Dunlin, doubtless a relic of Norn with the
later Scottish suffix -IE, but otherwise
unexplained.

Bottle Bump, see BUMBLE.

Bottle Jug. A Yorks. name for the
Long-tailed Tit. *Bottle* refers to the

shape of the nest, while *Jug* is a once popular pet form of *Jane*; see details under JUGGY WREN. Cf. JACK-IN-A-BOTTLE, JUG POT.

Bottle Nose, see GULDENHEAD.

Bottle Tit. A quite widespread local name for the Long-tailed Tit, *bottle* alluding to the shape of the nest, as in BOTTLE JUG. Cf. BARREL TIT.

Bouger, also spelt **Bowger.** A Hebridean term for the Puffin, making its debut in Martin 1698; the Gaelic spelling is *bùgair*.

A word of this shape may be analysed as root *bùg-* extended by *-air*, the latter a suffix productive in bird names rather like English -ER. But the root is more problematic, having no recognizable affinity with any other Gaelic word. Hebridean forms are, however, often ultimately Norse, and *bùg-* has a direct correspondence in Old Norse *būkr* (root *būk-*) belly. Unless appearances are deceptive, we have here a term reminiscent of BULKER, with the rider that the name would originally denote the corpulent nestling Shearwater.

Bramble. A term for the Brambling, first noticed in Merrett 1667 '*Montifringilla* Bramble, or Brambling', an entry copied by Ray 1678. We suspect a printing error, and that Merrett intended 'Bramble Finch' (to correspond to Latin *-fringilla* finch). Since BRAMBLING itself is a misnomer, the present word must in any case be spurious.

Bramble Finch. First noticed in 1865, a back formation from BRAMBLING, coined for systematic uniformity, FINCH representing scientific *Fringilla*; cf. MOUNTAIN FINCH. Though quite often used, it has not, of course, replaced BRAMBLING as the standard term. As a name, it is naturally as inapposite as its source. See also BRAMBLE.

Brambling. First occurring in Turner 1544 'Bramlyng', next 1570 'Bramlin', then 1655 'Bramblings', Skinner 1671 'Brambline', further Merrett 1667 '*Mon-tifringilla* Bramble, or Brambling', similarly Ray 1678, who also describes what he regarded as a separate species under the name 'The great pied Mountain-Finch or Bramlin: *Montifringilla . . . major*'. With Pennant 1776 'Brambling' the present word effectively became the standard term rather than MOUNTAIN FINCH. Inevitably, it has been assumed to derive from bramble, an etymology given as early as Skinner. Yet it is hard to envisage what this bird can have to do with brambles.

It will be noticed that in the earliest attestations the name appears as 'Bramlyng' and 'Bramlin', the latter deriving from the former; see -LING. Whereas these forms could theoretically represent colloquial pronunciations of a name Brambling, they could not improbably stand for Brandling. This latter word is widely found in local use, e.g. as a name for a young salmon or trout, a Brandling being a creature with branded or brinded, i.e. brindled, markings, and indeed both Bramling and Bramlin are among the colloquial forms recorded as fish names (*EDD*). Now this word would be equally suitable to name the bird we are considering, for its striking, brindled plumage is calculated to evoke a name, as in the local synonyms TARTAN BACK and FRENCH PIE FINCH. Quite evidently, then, the bird name too is properly **Brandling**, not Brambling, now seen to be a misnomer resulting from an erroneous normalization. Indeed, the correct form was known to Ray 1678 who, in spite of '. . . Bramlin: *Montifringilla major*' (above) occurring in his text, puts '*Montifringilla major* the greater Brandling' over the drawing of the bird at the back of the book (Tab. lxxvii). Should one not try to revive the correct form?

Brame, see BREEM.

Brandling, see BRAMBLING.

Brand Tail. A name for the Redstart recorded in Yorks., Derbys., Shrops., Worcs., and Heref., often written **Bran**

Tail in accordance with the casual pronunciation. Brand evidently means here fire-brand; cf. FIRE TAIL.

Brant, see BRENT GOOSE.

Bran Tail, see BRAND TAIL.

Branter. A Redstart name from Heref., based on BRAN TAIL; see -ER.

Brant Goose, see BRENT GOOSE.

Breem, also **Brame.** Terms for the Whimbrel locally in Suff. Whimbrel names fall into two classes: those which regard the bird as a kind of Curlew, e.g. HALF CURLEW, LITTLE WHAUP, and those which are based on the whistling, tittering cry, e.g. SEVEN WHISTLER, TIT-TEREL, or WHIMBREL itself. Noticing in Suff. dialect an imitative verb *breem* purr, we identify in Breem an artless onomatopoeic name, Brame being a secondary variant.

Brent. An abbreviation of BRENT GOOSE in use since Pennant 1768 'Mr. Willughby and Mr. Ray and M. Brisson very properly describe the *Bernacle* and *Brent* as different species.'

Brent Goose. Since Pennant 1768, for the earlier Ray 1678 'Brant Goose', first in 1597 'Foules, whom we call Barnakles, in the North of England, Brant geese' (*OED*). The earliest trace of the name is seen in the abbreviated (and Latinized) form in Turner 1544 'branta', the ultimate source being Old Norse *brandgás* literally burnt goose, where burnt is to be understood as (burnt) black; cf. BLACK GOOSE.

It appears that Pennant spelt the name Brent to accord with Gesner 1555 'brenta', Gesner for his part having 'corrected' Turner on the assumption that the word was to be connected with Greek *brénthos*, a waterbird (uniden-tified) mentioned by Aristotle. (Such an assumption is now known to be un-acceptable, if only because Germanic *b* corresponds to Greek *ph*; see the Intro-duction, pp.7–8.) In the event, Pennant's form found general acceptance, but the

traditional **Brant Goose** was still occa-sionally being used in the last century, as was also the abbreviated **Brant**.

As the citation from 1597 reminds us, the Brent and Barnacle were formerly not recognized as separate species, and we may add that local names are usually applicable to both; these include CLAIK, HORRA GOOSE, ROOD GOOSE, WARE GOOSE. See BARNACLE GOOSE.

Briskie. A Dumfries. and Kirkcud. term for a Chaffinch, alluding to its sprightly, brisk activity, -IE being the usual affec-tive suffix. Comparable motivation is found in DAPFINCH.

Broad Bill. A Lincs. name for the Shoveler.

Brook Ouzel. Occasionally quoted as a term for the Water Rail, apparently on the authority of Ray 1678 'The Water-Rail, called by some the Bilcock or Brook-Ouzel'. OUZEL is, however, hard-ly appropriate for a Water Rail. More-over, Hett 1912 finds the present name in local use for the Dipper, which makes entirely acceptable sense, cf. WATER OUZEL. We suspect that the record in Ray is due to some misunderstanding.

Brook Runner. A Worcs. name for the Water Rail, alluding to the stealthy running from one patch of thick reeds to another so characteristic of this bird, cf. the synonymous RUNNER.

Brown Owl. A name for the Tawny Owl, widespread locally, first recorded by Ray 1678 'The common brown or Ivy-Owl'.

Brown Starling, see BLACK STARLING.

Brown Swallow, see BLACK MARTIN.

Bud Bird. A term for the Bullfinch from Somerset and Devon, motivated by the bird's interest in the buds of fruit trees. Also **Bud Finch**.

Bulker. Given as a Hebridean name for the Puffin by Swainson 1885, evidently Gaelic **bulgair* literally (bird) with the *bulg-* belly. With this we may compare a local synonym *builgeun* literally belly bird (*builg-* = *bulg-*, *eun* bird). Such terms seem more appropriate to the bloated nestling Shearwater, which we take to have been the original meaning—as with PUFFIN itself. Cf. BOUGER.

Bull, see BULLFINCH.

Bullfinch. First in Turner 1544 'Bulfinche', apparently emerging as the recognized literary term in spite of several competing names, two of which are offered by Ray 1678 'Bulfinch, Alp, or Nope'. The first, however, was by then certainly the main term, since Ray 1674 himself characterizes the others as local or regional. Understandably Pennant 1768 'Bulfinch', since White 1789 usually in the present spelling.

There is a verse in the rhyme 'Who killed Cock Robin?' known from the first print of 1744: 'Who'll toll the bell?/ I, said the Bull,/ because I can pull,/ I'll toll the bell'. There can be little doubt that 'Bull' here means 'Bullfinch' (Opie 1952). Nor can there be much doubt that **Bull** is the original form of the name, as the parallel French *bœuf* literally ox, bullock, widely used in rural districts to denote this very bird, confirms. Although the motivation remains elusive, we are clearly dealing with nicknames. Given the impact of French upon English in the later Middle Ages, it seems probable that such an influence will ultimately explain the English name. In the light of the foregoing, Bullfinch is seen to be a secondary, rationalizing expansion of the original Bull. Cf. BULL SPINK; see also OX EYE.

Bull o' the Bog, see BUMBLE.

Bull Spink. A Yorks. name for the Bullfinch, first noticed in Merrett 1667 'Bull finch, Hoop, Bul Spink, Nope'.

Bully. A name for the Bullfinch, recorded from as far apart as Northumb. and Notts.; it has occasionally been used in literature. The earliest occurrence is traced to Cowper in verses headed 'On the Death of Mrs. Throckmorton's

Bulfinch' *c.*1788. It seems impossible to decide whether the present pet name was formed from the original BULL or from BULLFINCH or BULL SPINK.

Bum Barrel. A name for the Long-tailed Tit found locally from Lincs. to Wilts. *Bum* literally a swelling, something round, suitably qualifies barrel, here denoting the nest, seen as barrel-shaped. Popular names for this species often refer primarily to the unique nest, as BUSH OVEN, FEATHER POKE, HUCK-MUCK; cf. also BARREL TIT.

Bumble. A name for the Bittern, locality not quoted. The Old English name for this species is based on onomatopoeic *-dumbla*, etc., and this name type survives locally; see MIRE DRUMBLE. In the case of the present name, however, the initial *d* of the original form has been changed to *b* under the influence of BITTERN and its variants, which entered the language in the Middle English period. The name is not found in the earlier records; nevertheless, we surmise that it goes back to the late Middle Ages and we cite in support the homophonous verb from Chaucer (*Wife of Bath*) *c.*1386 'A bitore bumbleth in the mire'.

Somewhat later, an abbreviated *bump* makes its appearance as a noun and a verb: Skelton *c.*1508 'The bittern and his bumpe', Browne 1646 'A Bittor maketh that mugient [bellowing] noyse, or as we terme it Bumping', Dryden *a.*1700 'As a bittour bumps within a reed'. Such usage led to the creation of tautological names for the bird: **Bitter Bump,** found locally in the Midlands, and **Butter Bump,** widespread in the Midlands and the north; see BITTERN. Sporadically in the Midlands and more generally in the south, these are replaced by **Buttle Bump,** in some places becoming **Bottle Bump**.

Other developments are seen in Scotland, as **Bog Bumper,** the epithet *bog* doubtless invoked for alliteration: it is a relatively recent word in the language, coming in from Gaelic in the 16th cent.

Next, **Bog Blutter** (*blutter* noise, itself onomatopoeic) and finally, from the Borders, **Bull o' the Bog**, motivated by the stentorian call, likened to a bull's bellowing.

Nowadays the Bittern booms, hence **Boomer** (locality not specified).

Bum Towel. A term for the Long-tailed Tit from Devon and Somerset. To all appearances a local innovation after the pattern of the more widespread synonym BUM BARREL, but the element *towel*, evidently the product of folk etymology, has not been further explained.

Bunivochil. A Hebridean term for the Great Northern Diver, the first mention and the spelling in Martin 1703. It is Gaelic *buna-bhuachaill* literally dumpy herdsman, a name motivated by the shape of the bird and the wild, far-reaching call. Herdsmen call the cattle, and it was this bird whose loud cry gave rise to the legend immortalized in Kingsley's poem *The Sands of Dee*. A word of this type was doubtless a seaman's noa name in the first place. Cf. EMBER GOOSE.

Bunting. A name properly denoting the Corn Bunting, recorded since *c.*1300. It is, however, found as a surname a little earlier in a document of 1275. The literal meaning is plump or thick-set person or creature, an appropriate designation for this bird. Its streaky brown plumage lacks outstanding features, but its size, intermediate between a Sparrow and a Thrush, and particularly its dumpy form, cf. CORN DUMPLING, clearly identify it.

Pennant 1768 still uses the present name specifically, but at the same time refers to 'The Buntings'. Such generic use, equating the vernacular word with Linnaeus 1758 *Emberiza*, becomes definitive in Pennant 1773, whence the specific COMMON BUNTING, etc., of Pennant 1776. See also SNOW BUNTING.

Buntlin, i.e. **Buntling.** A Scottish form of BUNTING, with substitution of the common suffix -LING for the less usual -ING.

Burrian. A name for the Red-throated Diver, known from Ayrshire and Ulster. It reflects unrecorded Gaelic **buirrean* literally roarer, motivated by the stentorian voice.

Burrow Duck. A rather widespread local name for the Shelduck, first in Ray 1678 'Sheldrake, or Burrough-Duck, called by some, Bergander'. The bird often nests in a rabbit burrow.

Bush Oven. A Norf. name for the Long-tailed Tit, properly the intricate, dome-shaped nest, here likened to an oven. A word originally denoting the nest often features as a name for this Tit, see e.g. HUCKMUCK, FEATHER POKE.

Bustard. Virtually lost to England since it last bred here in 1832, this stately bird, specifically the Great Bustard, is first mentioned *c.*1460 'Bustarde'. However, the name must have been familiar before this, being recorded as a surname in 1391 (*OED*).

The etymological starting-point is the Latin of Pliny *c.*77 *avis tarda*, subsequently modified to **vistarda*, whence Italian *bistarda*, borrowed into French as *bistarde*. Meanwhile, on Gallic soil, *avis tarda* had evolved differently, changing via Late Latin **austarda*, a stage surviving unchanged in Provençal, into French *outarde*. It appears that both French names were brought to Britain, where they mixed to give the uniquely English Bustard—the form **Bistard**, found in some 16th- and 17th-cent. sources, e.g. Turner 1544 'Bustard, or Bistard', looks like a learned correction.

Avis tarda literally slow bird, is a strange name for a species so fleet of foot, and has understandably been regarded as a corruption due to folk etymology. One observes, however, that Latin idiom permits the pleonastic use of *avis* bird before the actual name, e.g. *avis strūthiō* as well as *strūthiō* ostrich. Pliny stated that the expression was Spanish. It is thus not improbable that *tarda* was, in fact, the original name, coming then from a language spoken in Spain before the arrival of the Romans. On being taken into Latin, it naturally seemed to mean slow. See GANT.

Butcher, Butcher Bird. A Shrike name, first in Merrett 1667 '*Lanius* Butcher, or murdering Bird', becoming in Wilkins 1668 '*Lanius*, or Butcher Bird'. This latter Ray 1674 uses as the literary word 'The great Butcher-bird, called in the Peak of Derbyshire wirrangle', similarly in Ray 1678. Pennant 1768 adopted it, too, but in later editions substituted SHRIKE.

Although Newton 1896 could state that Butcher Bird was the best known popular word for a Shrike, it is not, as we have seen, a genuine folk word, while Butcher itself is, of course, simply Merrett's translation of Latin *lanius*, a term introduced by Gesner 1555, obviously suggested by the bird's habit of impaling its prey on thorns.

Butter Bump, see BUMBLE.

Butterie, see BITTER BANK.

Buttle Bump, see BUMBLE.

Buzzard. First recorded *c*.1300 'busard', the present spelling dating from 1616. The name comes from French *busard*, known since the 12th cent., itself deriving with a suffix change from *buson*, older *buison*, which goes back to Late Latin **buteōnem*, properly accusative of the Classical *būteō*. The word is based on the onomatopoeic syllable *būt*, a representation of the shrill, long-drawn-out cry.

Locally, the term may be used for the Kite, but the early naturalists established the modern sense.

See PUTTOCK, KITE.

C

Cad Crow, see KET CROW.

Caddaw. An East Anglian name for the Jackdaw, first seen in *Prompt.Parv.* c.1440 'cadaw, or keo, or chowghe'. It is quoted by Turner 1544 '*Monedula* . . . a caddo, a chogh, a ka', but was not taken up by other naturalists. Locally also **Cadder;** see -ER.
　Could be spelt **Cad Daw,** i.e. DAW qualified by *cad* which in this area means carrion, as also in Lincs. **Cad Crow** (under KET CROW).

Caddess. A Yorks. term for the Jackdaw, formerly in literary use, the earliest citation being a dictionary entry of 1573 '*Monedula*, a chough, a daw, a cadesse'. From CADDAW under the influence of HAGGESS Magpie.

Caffincher, see CHAFFINCH.

Calaundre, Chalaundre. Old French forms occurring in English sources of the 14th and 15th cent. The French itself comes from Provençal *calandra* Calandra Lark, known since the 12th cent., but an ancient name, ultimately connected with Greek *kálandros*. This Mediterranean species was greatly esteemed for its fine, powerful song, and the name found its way into the medieval literature of countries even where the bird itself is unknown. See WOODLARK.

Calloo, also written **Caloo,** the pronunciation of the first syllable being *cal-*. The Orkney and Shetland name for the Long-tailed Duck, it echoes the musical call. Most popular terms for this noisy species are imitative, as COLDIE, DARCALL, also COAL-AND-CANDLE LIGHT.

Canada Goose. A typically North American species, described and so named by Ray 1678. It was introduced into Britain in the 18th cent. as an ornamental bird, the present wild population being founded by birds escaping from private collections.

Capercaillie. First spelt in this way in 1799, also since 1830 'Capercailzie' (where z=y), see -IE, for older 1536 'Capercailye', further the earliest attestation 1526 (misprinted) 'Auercalze'. The name is a corruption of Scottish Gaelic *capull coille* literally horse of (the) wood, the motivation being the shrill, frenzied climax of the male's nuptial song, for all the world like the whinnying of a horse. Comparable motivation is found in HORSE GOWK.
　English naturalists were slow to take up the Scottish name, generally preferring WOOD GROUSE, but Gould 1837 used it ('capercailzie'), after which Yarrell 1843 established the standard use of the word in its present spelling.

Carara. A Hebridean name for the Great Northern Diver, known since Martin 1703. It is Gaelic, imitative of the stentorian call. Cf. ARRAN HAWK.
　SGS, xii 23–5 (1971)

Cargoose, see CARR GOOSE.

Carner Crow, see CARRION CROW.

Carpenter Bird. A local name for the Great Tit, a development from a synonym of the type SAW SHARPENER.

Carr Goose. An old and apt name for the Great Crested Grebe, traditionally in use along the east coast, first attested in Charleton 1668 'Cargoos'. *Carr* is a local term for a marsh, deriving from Old Norse **kerr*. It is not unthinkable that this name actually arose as Old Norse **kerrgās*, which would take it back to Viking times. Cf. GAUNT.

Carrion Crow. First attested in 1528 'Carein Crowes . . . neuer medle with any quicke flesh'. The name appears in Merrett 1667 'common or Carrion Crow' as distinct from 'Royston Crow' (the

Hooded Crow). Both Ray 1678 and Pennant 1768 use the name in this way, thus *de facto* standardizing it. The word *carrion* came to us through the Normans; it goes back to Middle English where it is found in somewhat divergent forms, e.g. 14th cent. *caroigne, careyne*, borrowed without change from Old (Northern) French; the ultimate source is Late Latin (of Gaul) **carōnia*, a derivative of Classical *carō*, stem *carn*-flesh. The Norf. variant **Carner Crow** (see -ER) is a reminder of the erstwhile formal diversity of the word.

A word meaning carrion does not enter into Crow names in other Germanic languages. This type of name was, in fact, introduced from France, as indicated by the occurrence of the selfsame concept in provincial French (Anjou) *grolle charnière*. It follows that the alternative names, GORE CROW and KET CROW, though containing synonyms more ancient in English than *carrion*, will nonetheless not be older than the present term, but modelled on it.

In most parts of England the Carrion Crow is not traditionally distinguished from the Hooded Crow, nor is this surprising, given that the two sorts tend to occupy different areas. Only in some eastern districts where the Hooded is a winter migrant and the Carrion the resident bird, did the necessity for a special name for the visitor arise; see further under HOODED CROW, ROYSTON CROW. Otherwise CROW, pure and simple, is the sole name popularly required for either. It goes without saying that both birds are equally partial to carrion, so that one wonders how Carrion Crow came to be distinctively applied to just one of them. At this point we note that, in several places, the Rook is commonly regarded as a Crow and so named. It now becomes evident that Carrion Crow originally emphasized the difference between a Crow proper and a Rook, the modern contrast between the terms Carrion Crow on the one hand, and Hooded (or Royston) Crow on the other, being an arbitrary convention due to the early naturalists.

Castrel, see KESTREL.

Cataface. An Orkney name for an Owl. A relatively recent word as the purely English element *-face* shows, but *cata-* may be ultimately Norn; see further under KATOGLE.

Cat Owl, see KATOGLE.

Chack, see STONE CHACK.

Chaffinch. A traditional name going back to Old English *ceaffinc*; the bird would often be seen around the threshing floor or at the barn door searching for grains among the chaff. Despite numerous synonyms, such as DAPFINCH, PIE FINCH, SCOBBY, SPINK, the present word has for long been felt to be the chief name, at least since Turner 1544. A pet form **Chaffie** may occur locally. There is a variant with initial *c-*, not *ch-*, attested first in *Prompt. Parv.* *c.*1440 'caffynche' beside 'chaffynche', a second instance in 1570, to all appearances reflecting northern dialect *caff* chaff. Curiously, however, a comparable form has been recorded in modern times from the far south, from Sussex and Surrey, in **Caffincher** (where redundant -ER is of course secondary). See FINCH.

Chait. A Worcs. term for the Spotted Flycatcher, imitative of its usual call, a high-pitched, squeaky note.

Chalaundre, see CALAUNDRE.

Chalder. The Orkney name for the Oystercatcher, a Norn name from Old Norse *tjaldr*; cf. Shetland SHALDER. There is a variant **Chaldro** most likely traditional. A further form **Chaldrick** shows the diminutive ending -ICK of recent, Scottish provenance. It can be demonstrated philologically that *tjaldr* is of Germanic age, and from an echoic root *tel*, recalling the shrill, piping call of the breeding birds. See OLIVE, TIRMA. *Fróð.* xxiii 27–31 (1975).

Chan Chider. A Wilts. name for the Sedge Warbler, so called because it chides i.e. scolds so. The first word is obscure, but doubtless at least partly imitative, cf. CHAT, CHITTER CHAT.

Char Bob. A Derbys. name for the Chaffinch. The significance of *char* is not clear.

Char Cock. A northern name for the Mistle Thrush, in Westmor. appearing as **Cher Cock**, also **Jer Cock**, in Lancs. and Yorks. as **Sher Cock**. The first element of these obviously related names will ultimately represent the shrill, screeching flight call; cf. SCREECH, SHRILL COCK.

Chat. A name for the Sedge Warbler from the Thames Valley and Midlands, imitative of a characteristic call. This echoic word is also a traditional constituent of several other bird names, notably in the standard terms STONECHAT and WHIN-CHAT, where it has systematic value (=*Saxicola*) and hence may be used generically. Local names of other species include FALLOW CHAT, HEDGE CHAT, WALL CHAT (under WALL BIRD).

Chatter Pie. A name for the Magpie reported from places as far apart as Norf., Staffs., and Somerset. Comparable motivation is seen in TELL PIE.

Chauk. An expression in Devon and Cornwall for a Jackdaw, imitative of a note often heard from this bird, softer than the harsh cawing. It may be a relic of the former Celtic language of the area. See also JACKDAW.

Chawdaw. An Oxon. term for a Chaffinch, reminiscent of synonymous CHINK CHAWDY known from Cornwall, but both forms are unexplained. **Chawfinch**, in a record of 1611, is evidently a link with CHAFFINCH.

Check, see STONE CHECK under STONE CHACK.

Cheeper. A northern and Borders name for the Meadow Pipit, motivated by the plaintive call, as is commonly the case with Pipit names.

Chepster, see SHEPSTER under SHEP-STARE.

Cher Cock, see CHAR COCK.

Cherry Chopper. A Worcs. name for the Spotted Flycatcher, ostensibly prompted by an (unexplained) misconception that this bird attacks cherries. It seems likely that *cherry* is, in fact, corrupt. The same name type is found again in **Cherry Sucker**, originally **Fucker**, literally meaning cherry beater; for further details see WIND FUCKER, WIND SUCKER.

Chick. A shortened form of CHICKEN, occurring by the early 14th cent.; it is a constituent of the name DABCHICK.

Chicken. Used in the name MOTHER CAREY'S CHICKEN, the present word goes back to Old English *cicen*, *cycen*, an item of West Germanic age, cf. Dutch *kieken*, (Northern) German *Küken*, showing the diminutive suffix *-en* attached to a root presumed to be onomatopoeic, cf. COCK. (Southern) German *Küchlein* and Old Norse *kjúklingr* contain the same basic element, but with different diminutive suffixes.

Chickerel. A Dorset (Poole) name for the Whimbrel, imitative of the call, formed from the echoic verb *chicker* under the influence of the suffix -EREL.

Chiffchaff. First noticed in White *c.*1780 'The smallest uncrested willow wren or chiffchaf . . . utters two sharp piercing notes', doubtless the local name in Hants. It is, of course, expressive of the two notes referred to. The earlier naturalists often termed this bird 'Lesser Pettychaps', but Yarrell 1843 happily preferred the present folk-name, so establishing it in the standard nomenclature.

Chimney Swallow. A name for the Swallow, which, as Ray 1678 notes, may build in chimneys. Found locally in Wilts., it first occurs in White 1775 'The house swallow, or chimney swallow'. Pennant 1776 used the term, and it was taken up by Selby 1825, though never finding general acclaim, as referred to under HOUSE SWALLOW.

Chink. A Chaffinch name, found in the south from Essex to Cornwall, one of several onomatopoeic synonyms; see PINK. In Cornwall also **Chink Chawdy**; see CHAWDAW.

Chit. A Pipit name occurring locally in northern counties, being an abbreviated form of **Chit Lark**, recorded from Yorks. The literal sense is Little Lark, *chit* being a word for a small creature; see further under TIT LARK.

Chitter Chat. An onomatopoeic name for the Sedge Warbler, recorded from Northumb.; see CHAT.

Chitty Wren. A north country name for the Wren, *chitty* meaning tiny; see CHIT. Also **Chitter Wren**, i.e. with substitution of -ER for -Y. Cf. TITTY WREN.

Chough. A name going back to *c*.1305 'chogen' pl. (g=gh), then *c*.1381 'chowgh', *c*.1387 'choughe'. Evidence for the change of pronunciation from gutteral *gh* to *f* is found in Hodges 1643 and the word was occasionally so spelt: 1688 'The Jack Daw, or Daw . . . in some places called a Caddesse or Choff'. As the last quotation indicates, the present name could denote the Jackdaw, witness also Turner 1544 '*Monedula* . . . a caddo, a chogh, a ka' and Merrett 1667 'Jackdaw (or) Chough'; and it is beyond doubt that this is, in fact, the original meaning. There is no evidence for an English folk-name specifically for the Chough; it would never be prominent enough to be called anything other than Jackdaw or Crow.

The change of meaning came about as follows. The Chough (*Pyrrhocorax pyrrhocorax*) was once abundant in Cornwall and writers, from Turner 1544 on, often gave it the name **Cornish Chough** which for them, of course, was literally Cornish Jackdaw—a record of *c*.1575 speaks of 'Cornyssh daw', hence Sibbald 1684 called the bird 'Cornwall Kae'. This practice continued through Ray

1678 (but not Pennant, see RED-LEGGED CROW) to authors of the early 19th cent., as Fleming 1828, Selby 1833, Jenyns 1835. Meanwhile Bewick 1804 had dispensed with the epithet *Cornish*, Gould 1837 followed suit and Yarrell 1843 definitively established Chough, pure and simple, as the recognized vernacular equivalent of *Pyrrhocorax pyrrhocorax*. However, Cornish Chough remained an optional alternative, useful when Chough is employed generically, e.g. to distinguish the British bird from the exotic Alpine Chough (*P. graculus*).

Chough is an onomatopoeic name; it is the southern counterpart of northern KAE.

Chuck. A name for the Stonechat or Whinchat, cf. FURZE CHUCK, also for the Wheatear; see STONE CHUCK under STONE CHACK, also CONEY CHUCK. The name echoes the call.

Church Owl. A name first encountered in Ray 1678 'The common Barn-Owl, or White Owl, or Church-Owl', reported to be still in use in Wilts. This bird may use a church building as a nesting site.

Churn Owl. A very local, southern term for the Nightjar, first seen in Ray 1678 'The Fern-Owl, or Church-Owl, or Goatsucker'. This bird is often called Owl in folk-names, thus GOAT OWL, while the epithet derives from the churning song peculiar to the species.

Churre. A Norf. name for the Dunlin, onomatopoeic, with a local alternative PURRE.

Churr Muffit. In Scotland MUFFIT is a local Whitethroat name, *churr* being imitative of the harsh call note.

Cinereous Eagle. A term for the White-tailed Eagle coined by Pennant 1776; it was copied by several of his successors, last appearing in Jenyns 1835.

Cirl Bunting. Latham 1783 thus rendered Linnaeus 1758 *Emberiza cirlus* (see BUNTING), succeeding writers following suit. The term *cirlus*, due to Aldrovandi

1603, is a Latinized form of his Bolognese dialect *cirlo*, an imitative name, cf. *cirlar* (to) chirp, corresponding to Standard Italian *zirlare*. The Cirl Bunting is not a widely known bird in this country, being confined to the south-west and rather rare, indeed it was not scientifically recognized as a British species until 1800. Nevertheless, it appears to have acquired locally a folk name FRENCH YELLOWHAMMER, though this could never compete with the present term as far as standard usage was concerned.

Claik. A Scottish name for the Barnacle Goose, known since the mid-15th cent., imitative of the raucous call, as with most Goose names.

Clatter Dove. A Yorks. name for the Wood Pigeon, motivated by the loud clattering of the wings, which forms part of the male's courtship flight.

Cleaver. A local West Cornish name for the Manx Shearwater (Nance 1963), the literal sense being doubtless chopping axe, cf. HACKBOLT, an allusion to the characteristic flight pattern; see SHEARWATER.

Clee. A Redshank name, locality not given. It is imitative of one of the several calls, the bird having a considerable vocal range. Cf. TUKE.

Clinker. A Norf. name for the Avocet, inspired by the loud call, *clink* meaning to make a sharp, abrupt, ringing sound.

Clot Bird. An obsolete name for the Wheatear, first appearing in Turner 1544 '*caeruleo* a clotburd, a smatche, an arlyng, a steinchek' and occasionally found in later writers, the last record dating from 1753; some of these attestations, if indeed not all, are dependent on Turner. *Clot* is an earlier form of *clod*. On arable land, this bird's behavioural flitting from stone to stone may be modified to flitting from clod to clod, hence the present name.

Coal-and-candle-light. A fanciful name, Scottish and Northumbrian, for the Long-tailed Duck, clearly inspired by onomatopoeia; see CALLOO.

Coal Goose. A Kent term for the Cormorant, evidently motivated by the black winter plumage.

Coal Hood. A north country and Scottish name applied to various small birds with black heads, as the Bullfinch, the Coal Tit, and the Reed Bunting.

Coalmouse, also spelt **Colemouse.** An obsolete name, originally loosely applied to various Tits, i.e. not only the Coal Tit, but also the Great and the Marsh (and Willow) Tits. It goes back via Middle English *colmose* to Old English *colmāse*, consisting of *col* coal, which naturally refers to the black plumage of head and throat so noticeable in the species mentioned, and *māse*, an ancient word, itself having the general sense of TIT; see further under TITMOUSE. The present name is of West Germanic age, witness Dutch *koolmees* and German *Kohlmeise*, which in the modern standard languages have acquired the specific sense of Great Tit. Pennant 1768 used 'Cole Mouse' specifically to denote the Coal Tit, perhaps feeling that this vernacular name best matched Gesner 1555 *Parus ater (āter* black), but altered it in 1776 (following Tunstall 1771) to 'Cole Titmouse', clearly for systematic reasons (*Parus* = Titmouse); see COAL TIT.

The spelling Coalmouse is first seen in 1829 'coal or colemouse'.

Coal Tit, formerly **Cole Tit.** Generalized by Yarrell 1843 'Cole Tit', for earlier COAL or COLE TITMOUSE, first in Tunstall 1771 'Cole Titmouse'; see further under COALMOUSE. The spelling *coal* is first noticed in Bewick 1826 'Coal Titmouse'; it becomes usual by the end of the century (BOU 1883 'Cole Titmouse', Hartert 1912 'Coal Titmouse'). See also TIT.

Coatham Crow. A local Cleveland name for the Hooded Crow, a winter migrant to this district, here as elsewhere often seen on the seashore. The reference is to Coatham, a seaside settlement (now part of Redcar). Similar name-giving is found in SCREMESTON CROW, also ROYSTON CROW.

Cob. A term for a Sea Gull, known since 1580, also as **Sea Cob** since 1530, in use at various places on the east and south coasts. Apparently a very old name, fortuitously absent from the earliest records, witness synonymous Dutch *kobbe*. Further connections are uncertain.

Cobbler's Awl. A Norf. term for the Avocet, alluding to the long, thin curved beak.

Cock. A constituent of numerous specific names, going back to Old English *cocc*, recorded as early as 897 and evidently already present in the language of the Anglo-Saxon invaders, since it occurs, in Latin dress, as *coccus* in the Franconian *Lex Salica* of the late 5th cent. Clearly echoic in origin, it competed during the Old English period with a still older name *hana* (see HEN), entirely replacing it by the Middle English period, when it would be reinforced by French *coq*, of Germanic origin, introduced by the Normans.

Altogether Cock has been very productive in bird names, but losing its original sense to become the equivalent of bird; in such use it may be particularly associated with a given species, e.g. in the Water Rail names BILCOCK, DARCOCK, SKITTY COCK.

Cock can also mean Woodcock, its employment in this sense being traceable as far back as 1530, the date of the first recording of *Cockshoot* 'a broad way or glade in a wood, through which woodcocks, etc., might dart or "shoot"; so as to be caught by nets stretched out across the opening' (*OED*). It is one of those vague fowler's terms and, like WOODCOCK itself, was most probably evasive in the first place.

Used attributively, Cock naturally refers primarily to the male, but in

several instances this proper sense is not (or no longer) tangible, the word being a meaningless pendant, or at the most adding a touch of familiarity, as COCK FELT, COCK ROBIN, COCK SPARROW, COCK WINDER. None of these terms is known to be older than the end of the 16th cent. The other type, seen in BILCOCK, etc., is older: a Middle English example is quoted under THRICE COCK.

Cockathodon, also **Cockathrodon.** A Scillonian name for the Manx Shearwater, making its debut in Yarrell 1843; the variant with -r- first occurs in a manuscript of 1850. Locally, the name is interpreted as imitative of the insistent call of the breeding birds. On the other hand, a word of this phonetic shape, stressed as it is on the third syllable, argues for an underlying Celtic compound name. In that case, the forms of the name current today could conceivably be corruptions of Late Cornish *cok an dodn* literally cock of the wave, a not unacceptable name for this species. *TPS* 173f. (1975)

(Cockerel. Known since *Prompt.Parv.* c.1440. It consists of COCK and the suffix -EREL in a diminutive function, with DOTTEREL the earliest recorded example of the use of this suffix in connection with terms denoting birds. Though not a specific bird name, the mere existence of such a common word as the present may well have encouraged the adoption of the distinctive -EREL found in the names of certain species.)

Cock Felt. A Northants. name for the Fieldfare; see COCK, FELT.

Cock of the Wood. A name for the Capercaillie, first noticed in 1649 as 'A bird peculiar to Ireland'. Quoted by Ray 1678, it obtained a certain currency in older ornithological works. As the syntactic form indicates, the name is of Anglo-Irish provenance, inspired by Irish Gaelic *coileach feá* (*coileach* cock, *feá* of wood)—it properly denoted the Irish race, which became extinct in the latter half of the 18th cent.

Cock Robin. The name makes its debut *c.*1699 as a cant expression for a 'soft, easy fellow' (*OED*). In reference to the bird, the earliest occurrence is seen in the rhyme 'Who killed Cock Robin' in the first print of *c.*1744. It seems impossible to say exactly how *cock* arose in the present connection. Since the sexes are identical in appearance, one could interpret *cock* as a familiar, but sexless nickname. On the other hand, folk tradition has it that the Robin is a male bird, the husband of the Wren.

Cock Sparrow. A familiar term for the House Sparrow, known since 1598. See COCK.

Cock Winder, see WINDER.

Cold Finch, Cole Finch. Names for the Pied Flycatcher, the former traceable to a statement in Ray 1678 'A bird called a Coldfinch by the Germans'. But no such German word is known. According to Swainson 1885, the bird had a north country name **Cole Finch**, i.e. black finch, a very suitable name for this species. It rather looks as though Ray's Cold Finch is a garbled form of this name. Edwards 1743 uncritically followed Ray, as did Pennant 1768, but Pennant 1776 substituted PIED FLYCATCHER, thereafter the standard name.

Coldie. A Forfar name for the Long-tailed Duck, consisting of the basically imitative root *cold* (cf. CALLOO) with the diminutive ending -IE.

Colemouse, Cole Tit, see COALMOUSE, COAL TIT.

Colk. A name for the Eider used by early writers in their accounts of the Western Isles of Scotland, the earliest dating from 1549. The name is Gaelic *colc,* explicable as meaning down, the word apparently having been extracted from (obsolete) *colcaidh* feather-bed. The present name has thus the same basic sense as EIDER itself.
    SGS x 58f. (1963)

Collared Dove, Collared Turtle Dove. After Bonaparte had, in 1855, instituted the genus *Streptopelia,* coined from Greek *strepto-* collared and *péleia* dove, with reference to a stripe on the side or back of the neck, the new term could readily be Anglicized as Collared Dove or, more distinctively, Collared Turtle Dove. In practice, however, these properly generic names became attached to different species or subspecies. The latter tended to refer to *S. decaocto,* the former to *S. roseogrisea,* a difference of meaning still observed in Hutton 1956. But such terminological niceties were upset by the dramatic colonization of Britain by *S. decaocto,* officially since 1955. Now the bird was constantly being talked and written about. By the 1960s the rather long-winded Collared Turtle Dove had generally been abbreviated to Collared Dove, regardless of previous usage.

Collier. A Yorks. word for the Swift, in humorous reference to the dark plumage; see further under BLACK MARTIN.

Colly, also **Colly Bird.** Terms for the Blackbird from Glos., Somerset, and Devon. The former is an abbreviation of the latter, *colly* being a form of *coaly,* i.e. black.

(Common. An epithet first noticed in Merrett 1667 'common or carrion crow', etc. It was used by later naturalists, eventually becoming a frequent constituent of specific names. Examples below.)

Common Bunting. A name for the Corn Bunting, proposed by Pennant 1776 in place of the earlier BUNTING which had been given generic status in Pennant 1773. The new name was accepted by his successors, including Yarrell 1843 and subsequent editions, but BOU 1883 preferred CORN BUNTING, after which the present term fell more and more into disuse.

Common Buzzard. Introduced by Ray 1678 'the common Buzzard or Puttock'.

It was taken up by Pennant 1768 and is still the technical equivalent of *Buteo buteo*, but British bird books nowadays mostly ignore the epithet.

Common Creeper. A term for the Tree Creeper coined by Albin 1738; it was taken up by most early naturalists (but not Pennant; see CREEPER) and was used by Yarrell 1843 and later editions.

Common Eider, see EIDER.

Common Gallinule, see MOORHEN.

Common Godwit, see BAR-TAILED GODWIT.

Common Grosbeak. Coined by Ray 1678 'The common Gros-beak or Hawfinch' and taken up by Pennant 1768 and most of his successors until replaced by HAWFINCH.

Common Guillemot. Introduced by Yarrell 1843, who perhaps felt that FOOLISH GUILLEMOT, used by his predecessors, was too frivolous for scientific use. At any rate the new term at once established itself as the ornithological standard.

Common Gull. Coined by Pennant 1768, who described it as the most numerous of the genus, Ray 1678 'common Sea-Mall' providing the prototype.

Common Partridge. A term introduced by Ray 1678 for systematic reasons, the bird's usual name being simply PARTRIDGE, at the time the only species found in Britain.

Common Rotche, see ROTCHE.

Common Sandpiper, see SANDPIPER.

Common Scoter. A term introduced by Yarrell 1843 which thereupon replaced BLACK SCOTER.

Common Sheldrake. Introduced by Yarrell 1843 and used in BOU 1883; see SHELDUCK.

Common Skua. A name for the Great Skua coined by Fleming 1828; see also SKUA. The name was adopted by succeeding writers. It appears in BOU 1883, but was not long afterwards replaced by the more appropriate term GREAT SKUA.

Common Snipe. A term brought in by Pennant 1768 for systematic reasons, the species being previously known simply as SNIPE, as it still is colloquially.

Common Tern. First in Pennant 1812 beside 'Great(er) Tern' of earlier editions. The choice of the epithet *common* was at the time justifiable, since the Arctic and Roseate Terns had not yet been identified as species separate from this bird.

Common Whitethroat. A term first used by Latham 1787 to distinguish this species formally from the then recently discovered LESSER WHITETHROAT. But today, at least in ordinary use, the present bird is simply WHITETHROAT.

Coney Chuck. A Norf. name for the Wheatear, which may nest in a coney's (rabbit's) burrow. The onomatopoeic CHUCK also occurs in the synonymous STONE CHUCK under STONE CHACK.

Cooscot, Cooshot, see CUSHAT.

Coot. The name goes back to Middle English *cōte*, first attested *c*.1300 'cote'. The name was used by the pioneer naturalists, e.g. Turner 1544 'cout', Merrett 1667 'coot', and automatically became the standard name for the species. Secondary variants occur, reflecting dialect changes, the earliest examples being *c*.1450 'Beld cyttes' (see BELL KITE) and *c*.1475 'cute, cuytt' (see CUTE, QUEET).

The word is onomatopoeic, echoing the high-pitched call. It has a formal parallel in Dutch *koet*, known since the 16th cent., suggesting a common origin in West Germanic *kōt-*, fortuitously unrecorded in the oldest monuments of the languages concerned. Imitative names, however, can easily arise at any time, so that the possibility of the English and Dutch being relatively late, independent creations is not excluded. In Old English an entirely different word *dopened* literally dipping duck (see ENDE) is equated with Latin *fulica* coot.

Corbie. A Scottish name for the Raven, occasionally extended to the Crows and the Rook, known since c.1420. It is a diminutive form (see -IE) of Middle English *corb*, a loan from Old French **corb*, a variant of (the actually recorded form) *corp*; the name goes back to Latin *corvus*.

Cormorant. The name makes its entry c.1320 'cormerant', next Chaucer c.1381 'cormeraunt', the modern spelling first in Turner 1544, similarly Ray 1678. Pennant 1768 altered this to 'Corvorant', writing 'The learned Dr Kay, or Caius, derives the word . . . from *Corvus vorans*, from whence corruptly our word *cormorant*'. The fanciful emendation was, however, not generally followed.

The source of the word is French, the oldest known form being 12th cent. 'cormareg' for 'cormareng' from *corp* raven and **marenc* marine, cf. 8th cent. Latin *corvus marīnus*. The final *t* of the English form is secondary, as also in PHEASANT. See CORBIE. Cf. GOR MAW.

The present term has replaced a native name seen in Old English *scealfor* corresponding to Dutch and Low German *scholver* cormorant.

Local names include HIBLIN and HUP-LIN, LAIRBLADE, LORIN, PARSON.

Corn Bunting. Although not recorded until Montagu 1831, who describes it as provincial, this name is to all appearances traditional. The older ornithologists used the artificial term COMMON BUNTING, but BOU 1883 gave its backing to the present name, since which time it may be regarded as the standard.

Corncrake. The earliest attestations date from a.1455 'Corne Crake' and 1552 'Cornecraik'. Originally Scottish, the name entered English ornithological literature through Bewick 1797. It was

naturally chosen by the Scot Fleming 1828. Yarrell 1843 acknowledged the name by admitting it as an alternative to his main term LAND RAIL. After this the two competed, the present term proving the stronger; it had the backing of BOU 1883 and Hartert 1912. Since the 1920s it has become the generally recognized standard.

The name CRAKE is of echoic origin; see this word. The epithet is self-explanatory: the bird so often nested in the corn fields; cf. CORN DRAKE.

Corn Drake. A Yorks. name for the Corn Crake; see DRAKE 2.

Corn Dumpling. A picturesque name from Ulster for the Corn Bunting (Braidwood 1965), alluding to the robust form.

Corn Goose. A name for the Bean Goose, locality not specified; it is appropriate since the birds frequent cultivated fields, when they may take grain.

Cornish Chough, see CHOUGH.

Cornwillen. A name in Cornwall for the Lapwing, borrowed from Middle Cornish **cornwhylen*, consisting of *corn* horn (the crest being compared to a horn, cf. HORN PIE, HORNY WICK) and an obscure second element closely paralleled in the Welsh synonym *cornchwiglen*. The borrowing into English must have taken place before 1600 at the latest, since in more recent Cornish the word changed by regular phonetic evolution to *codnahwylan*.

The occurrence of substantially the same name in both Celtic languages indicates that an ancestral form was already present before the Celts of the south-west peninsula were cut off from those in Wales at the end of the 6th cent.

Coulter Neb. A Northumb. name for the Puffin, literally meaning plough-share bill, first recorded by Ray 1678 'called *Coulterneb* at the Farn Islands'. It has every appearance of being traditional, since both *coulter* and NEB are known to go back to Old English times.

Covie, Covie Duck. Northumb. terms for the Scaup. Unexplained.

Cow Bird, Cow Clit. Local names for the Yellow Wagtail. Unlike its cousins, this Wagtail is often seen among cattle, snapping at the insects disturbed by the grazing animals, hence the term Cow Bird from Norf. and Worcs. In the other name, from an unspecified locality, the element *clit* is imitative of the sharp call-note, suitably alliterating with *cow*.

Cow Prise. Given by Swainson 1885 as a north country name for the Wood Pigeon. One can only surmise that this obscure expression (if not spurious) developed from the name type illustrated by COW SCOT, etc.

Cow Scot, Cow Shot, see CUSHAT.

Cracker. A name for the Pintail known from Ray 1678 'Sea-Pheasant or Cracker', motivated by the alarm call heard as 'crack'; see -ER.

Cracky. A Devon name for the Wren, a term denoting a small creature; there is a dialect variant **Crackit.** Also **Cracky Wren.** Cf. STUMPIT.

Craff. A Cumb. name for the House Sparrow. Origin unknown.

Crag Ouzel, see ROCK OUZEL.

Craigit Heron. A Scottish name for the Heron, the epithet short for *lang-craigit* long-necked; also **Craigie Heron** with the diminutive of *craig* neck; see -IE.

Crake 1. A common local name for the Corn Crake, known since 1791, but of course much older, since it is a constituent of CORNCRAKE, the records of which go back to the 15th cent. Cf. also CRAKE GALLINULE.

The word is clearly imitative of the monotonous, rasping croak, as in other names for this bird; see DRAKE 2, RAIL.

Crake 2. A northern term for a Crow, first found c.1320. It is a relic of Old Norse *krāka*, the ordinary word for Crow in that language, echoic in origin and to that extent comparable to CRAKE 1. See CROW.

Crake Gallinule. A term for the Corn Crake coined by Pennant 1776 for systematic reasons, CRAKE being, of course, a well-known name for the species. See further under GALLINULE, also LAND RAIL.

Crane. Goes back to Old English *cran*, comparable to Dutch *kraan* and German dialect *Kran*. The name of this striking species is of Indo-European age, cf. Welsh *garan*, Greek *géranos*, also with change of meaning Lithuanian *garnys* heron, stork. The name is imitative of the trumpeting call and thus basically akin to Latin *grūs* crane.

This ancient word has been kept alive not only because the Crane itself remained a familiar bird, at least in Lincs., Cambs., and East Anglia, until the end of the 17th cent., but also because, in many other districts, the name was transferred to the Heron.

Craw. A northern and especially Scottish variant of CROW.

Creak Mouse. A Glos. term for the Long-tailed Tit (Jackson 1968), *creak* presumably from the sharp call; on *mouse* literally tit see TITMOUSE.

Creeper. An old name for the Tree Creeper, first encountered in Turner 1544 'creperam' (Latinized), further Ray 1674 'Creeper, or Ox-eye Creeper', the same in 1678, Pennant 1768 'creeper'. It was used again by Bewick 1797, but other early authorities preferred COMMON CREEPER.

Cress Hawk. A local name in Devon and Cornwall for the Kestrel, the first element being to all appearances a corruption of Old French *cresselle* kestrel; cf. CRISTEL HAWK, and see further under KESTREL.

Crested Tit. Generalized by Yarrell 1843,

for the older **Crested Titmouse**, first noticed in Charleton 1668 '*Parus cristatus*, the Crested or Juniper Titmouse'. The name is a translation of the Latin which derives from Gesner 1555. Confined to pine-woods in the Scottish Highlands, where the language was traditionally Gaelic, this species never came by an English folk-name.

Crick. An East Anglian term for the Garganey, imitative of the high-pitched rattle heard from the courting drake, and to all appearances traditional; cf. Low German *Krick* (also transferred to the Teal, hence Standard German *Krickente* in this sense).

Cricket Bird. A Norf. name for the Grasshopper Warbler, motivated by the whirring song, here likened to the chirping of the cricket.

Cricket Teal. An East Anglian expression for the Garganey, first seen in Merrett 1667 misprinted 'Crickaleel'. The name alludes to the characteristic rattle (see CRICK), here compared to the sound heard from the cricket. Cf. the synonymous SUMMER TEAL.

Cristel Hawk. A name for the Kestrel, locality not given. Apparently a corruption of Old French *cresselle* kestrel; cf. CRESS HAWK and see further under KESTREL.

Crocard. A name of an unidentified species known from a record dated 1541-2 'Crocards and Oliffs', but to all appearances a variant of CROCKER, the suffixes -ARD and -ER being interchangeable.

Crocker. Given as a name in Sussex for the Brent Goose, in Kent for the Blackheaded Gull. As a term for the Brent, it is plausibly formed from a syllable *crock*, representing the call, with the agent ending -ER, but this can hardly be the explanation of a name properly denoting a Gull. Cf. CROCARD.

Crooked Bill. A name for the Avocet, first recorded in 1708.

Crossbeak, Crossbill. After Charleton 1668 'Cross-beak', we find Ray 1678 'Shell-apple or Cross-bill', Pennant 1768 taking the latter, which remained as the standard term.

It is perhaps noteworthy that Merrett 1667 has only 'Shell-Apple', and seeing that the Crossbill is a little-known species in this country, it can hardly ever have had any widespread folk-name (see further under SHELL APPLE). Crossbeak and Crossbill will be terms coined by naturalists, ultimately reflecting Gesner 1555 *Loxia*, created by him from Greek *loxós* crossed, doubtless inspired by his native Swiss German *Krützvogel* literally cross bird.

Crotch Tail. An Essex name for the Kite, literally meaning Fork Tail.

Croupy Craw. A north country name for the Raven, the epithet referring to the crouping, i.e. croaking, call.

Crow. This word goes back to Middle and Old English *cráwe* with close correspondences in Dutch *kraai* and German *Krähe*. It appears that these reflect three preliterary variants, respectively **kráwa*, **krája*, **kráha*, the differing consonants being glide sounds independently developing between the vowels of a common **kráa*, implying a West Germanic feminine **kráō* literally crower, derived from **kraan* to crow (Old English *cráwan*, Dutch *kraaien*, German *krähen*), the verb itself, of course, of onomatopoeic origin; for the manner of formation, cf. *-gale* in NIGHTINGALE. The name is absent from North Germanic, its place being taken by Old Norse *kráka*; see CRAKE 2.

As explained above (Introduction, pp. 8–10), these words must have originally referred indifferently both to the Crows and the Rook. In popular use, CROW is still often so applied, occasionally including the Raven; see CROUPY CRAW; it may also be extended to other species, notably the Dipper; see WATER CROW.

Cuckoo, formerly **Cuckow.** The name first appears *c*.1240 'Sumer is icumen in/ Lhude sing cuccu!'. Subsequent occurrences include Chaucer c.1381 'cokkow', other manuscript readings 'cucko, kukkowe, cuccow, cuckow'; of these the last became the most usual, surviving well into the 19th cent. and which Newton 1896 sought to revive. The common 17th cent. spelling 'cuckoe', seen e.g. in Merrett 1667 'Cuckoe or Guckoe', where *oe* would be pronounced as in shoe, emphasizes the imitative character of the name. It prefigures the spelling Cuckoo, which came to the fore in the 18th cent., and was adopted by Pennant 1768 and most of his successors, Ray 1678 still having 'Cuckow'. Cf. GUCKOO.

The name was brought to this country by the Normans; it is Old French *cucu*, known since the 12th cent. This onomatopoeic representation of the two notes of the bird's call has replaced the native one-note representation seen in Old English *gēac*, Middle English *yek* which survived sporadically until the 15th cent. In the North, however, this traditional English name was replaced by its Norse cognate; see GOWK. But the present name was clearly felt to be the main term by the 17th cent., witness Holme 1688 'The Cuckow is in some parts of England called a Gouke.'

Cuckoo's Marrow. A Midlands term for the Wryneck, *marrow* being a local word (known since c.1440) for mate, in allusion to the bird's arriving a little before the Cuckoo.

Cudbert Duck, see CUTHBERT DUCK.

Cuddy. Essentially a term of endearment applicable to any small creature, hence in Yorks. used of the Hedge Sparrow, in Northants. of the Tree Creeper, in Somerset of the Wren, the latter not far removed from CUTTY.

Culbert Duck, see CUTHBERT DUCK.

Culver. This name for a Pigeon, found in many parts of the country and until this century much used in literature, goes back to Old English *culfer, culfre*, oldest

form *culufre*. The name derives from Late Latin *columbula*, a diminutive of Classical *columba* pigeon, in the spoken language contracted to *columbla*, then by dissimilation of the second *l* changing to *columbra*, both stages being attested in modern Romance dialects. The last form can be envisaged as further changing to **colubra*, which in the Latin of Gaul would regularly become **coluvra*, the immediate source of the English borrowing.

Among the symbols of Christianity, *columba* has been particularly prominent and, as a translation, the still rather similar-sounding *culufre* must have recommended itself to the Anglo-Saxon clergy, for this is the term used in the religious texts of the period, possibly as a noa word, to the exclusion of the inherited Germanic name **dūfe*; see DOVE. The present name is not applicable to the Turtle Dove; see TURTLE.

Culver Duck, Culvert Duck, see CUTHBERT DUCK.

Curlew. Known since the 14th cent., when we find a.1340 'curlu', 1377 'corlue', 1387 'curlewes'. It is a loan from Old French *corlieu* courier, but basically imitative of the wild, far-reaching call. It is the name generally used by the early naturalists, including the influential Pennant 1768, and thus established itself without difficulty as the standard term. A native English name survives in WHAUP; see also CURLEW HELP. A Scandinavian name is present in SPOWE.

Curlew Help. A Lancs. name for the Curlew. It is tautological, for there can be little doubt that *Help* continues, corruptly, the traditional English name for the Curlew, as explained under WHAUP.

Curlew Sandpiper. Selby 1833 coined **Curlew Tringa** for systematic reasons in preference to PIGMY CURLEW, introduced in 1785. Selby's English-Latin hybrid was then replaced by the present name in Yarrell 1843. He was not entirely without precedent, Bewick

1826 having called this bird PIGMY SANDPIPER. This species seems to have been without a folk name.

Curre. A name applied indifferently to the Goldeneye, Pochard, Scaup, and Tufted Duck, imitative of the harsh, growling call common to these diving Ducks.

Curwillet. First in Ray 1674 and again in 1678 'Sanderling, called also Curwillet about *Pensance* in *Cornwal*'. Pennant 1768 referred to the name, so that this otherwise obscure word gained a certain currency in ornithological works. A local West Cornish name of this age will have been Celtic in the first place, indeed at the time of Ray's visit Cornish, though fast declining, was still a living tongue in the Penzance district. The name has every appearance of being onomatopoeic, reproducing in some sort the shrill call uttered by the bird, especially when disturbed.

Cushat. A now mainly non-literary name for the Wood Pigeon, chiefly in the north and especially in Scotland. Indeed, the present spelling is Scottish (Burns, Scott); other local forms include widespread **Cow Shot**, Yorks. **Coo Shot** (of which Cushat is simply a variant) Yorks. also **Cow Shot**, **Coo Scot**. These types go back via Middle English *coushote, couscote* to Old English *cūsceote, cūscote*, with a very early variant *a*.700 '*palumbes* cuscutan' (pl.).

The forms with *cow* show the usual phonetic development from Old English *cū-*, those with *coo* reflect a dialect pronunciation which kept the Old English unchanged. Now whereas Old English *cū* meant cow, this can hardly be the primary sense here, which must be as one with cooing. In the second part, differentiation into the two types *scot* and *shot* began in the Old English period, but as neither is comprehensible as a bird name, we conclude that both are corrupt. On the other hand, the first attestation *-scutan* can be understood as **scūtan* shouters. We therefore normalize as **cūscūta* (sg.), giving the literal sense coo-shouter, i.e. in modern idiom coo-caller, a plausible name for the species in question. See QUEECE.

Cut. A dialect word meaning short, erect tail, used locally in the south to denote the Wren, hence tautologically **Cutty Wren**, a very widespread term, in some places shortened to **Cutty**. Cf. SCUT.

Cute. A Norf. variant of COOT.

Cuthbert Duck. A Northumb. term for the Eider. The bird breeds in the Farne Islands and the down collected from the nests gave this Duck especial importance, which led to its being placed under the patronage of St. Cuthbert (Bishop of Lindisfarne, died 687). The name is first recorded in Ray 1674, but the tradition is much older, as we learn from Reginald of Durham who *c*.1165 refers to the Eiders as being called, in addition to their ordinary name, '*aves . . . Beati Cuthberti*', birds of the Blessed Cuthbert. Clearly, too, this designation was at the time, or had at least arisen, as a secondary noa term, a reflection on the value of the down-producing bird.

Locally, the name *Cuthbert* was modified to *Cudbert*, hence **Cudbert Duck**, from this a variant **Culbert Duck**, further corrupted to **Culvert Duck**, and finally to **Culver Duck**.

Cutty, Cutty Wren, see CUT.

D

Dabchick. A fairly widespread local name, since Yarrell 1843 also in common use as an alternative to LITTLE GREBE, the element -CHICK naturally alluding to the bird's smallness. Making its debut in Merrett 1667, the present form derives through corrupt association with the living word *dab* from **Dapchick**, today a local Somerset word known since *a*.1575. As shown by the evolution of DIDAPPER, a still more archaic form is preserved in obsolete **Dopchick**, recorded in 1615 and 1732, also 1583 'Dopchicken', whence under the influence of Dabchick the Suff. term **Dobchick**, first noted by Ray 1678.

Dadfinch, Daffinch, see DAPFINCH.

Daker. A Surrey name for the Corn Crake, better known as **Daker Hen**. The latter occurs in various parts of the country, being first recorded in 1552. Corn Crake names are characteristically onomatopoeic. DAKER can thus be envisaged as an agent noun based on **dake*, imitative of the call, and close to synonymous DRAKE 2.

Dapfinch. In Devon a name for the Chaffinch, *dap-* reflecting the local verb *dap* hop, move briskly. This element may be assimilated, hence the variant **Daffinch**, or dissimilated, giving **Dadfinch**. Similar motivation is seen in the Scottish synonym BRISKIE.

Dapchick, see DABCHICK.

Darcall. A name for the Long-tailed Duck, locality not specified. The word is evidently onomatopoeic; cf. CALLOO.

Darcock. Given as a local name for the Water Rail, properly **Dartcock*, in reference to its characteristic darting from one piece of thick cover to the next, a habit which has often led to comparable names; see under RUNNER.

Darr. A name for a Tern, current in Norf.

and Suff., imitative of the harsh call, as usual with names of this species; see TERN.

Dartford Warbler. A name coined by Pennant 1776 who writes 'A pair were shot on a common near *Dartford*, in April 1773, and communicated to me by Mr. Latham.' See FURZE WREN.

Daw. Known since 1432–50 'dawe', since the 16th cent. yielding more and more to JACKDAW. To all appearances traditional and cognate with Old High German *taha*, pre-Old High German **daha*, basically onomatopoeic; cf. KAE.

Devil 1. A now forgotten name for the Coot found in a record of 1580 'a bird called a Coute, and because of its blackenesse, is called a Diuell'.

Devil 2. A Berks. name for the Swift, motivated basically by the dusky plumage which distinguishes it so conspicuously from the otherwise rather similar Swallow and Martins; cf. DEVIL 1 and such synonyms as BLACK SWALLOW, BLACK MARTIN. The concept is seen in local names all over the country. Thus one finds Yorks. **Devil Bird**, Suff. **Devil Dick** and (district not mentioned) DEVIL SWALLOW; see also DEVIL'S BITCH. From Wilts., Somerset, and Devon comes **Scare Devil**, locally often pronounced **Skeer Devil** and so written, from Northumb. **Swing Devil**, the epithet here having the sense of rush, whirl, in allusion to those headlong chases round the nesting sites. Devil is often combined with names based on the penetrating cry; see DEVIL SCREECH. One imagines that the precipitate flight of these sombre-feathered birds, and the incessant squealing which accompanies it, have served to make Swifts appear quite uncanny. In this connection there are mild examples of name taboo in the Suff. names DEVILTON and JACK-A-DELLS; see also DEVILING.

Deviling. A name for the Swift reported from various parts of the country from Westmor. to Suff. It stands for *Deviling literally little devil, a derivative of DEVIL 2, formed with the diminutive suffix -LING, here giving a euphemistic nuance, cf. DEVILTON, JACK-A-DELLS, further DICKY DEVLIN.

Devil's Bitch. A Yorks. name for the Swift, derogatory in keeping with the countrywide view of this bird; see DEVIL 2.

Devil Screech, also **Devil Screecher.** These fairly widespread local terms for the Swift combine reference to the devil (see DEVIL 2) with names denoting the piercing call, as do numerous other synonyms, including Hants. and Yorks. **Devil Screamer** and Yorks. **Devil Shrieker** or **Squealer,** the same doubtless in **Devil Screw** (see SCREW). Cf. SCREECH MARTIN.

Devil Swallow. A local name for the Swift; see DEVIL 2. Swifts are often known as Swallows; see SWALLOW.

Devilton. This Suff. name for the Swift, sounding more like a surname, as Hamilton, will therefore be best explained as a euphemistic derivative of DEVIL 2. Cf. JACK-A-DELLS, DEVILING.

Dicky Devlin. A Yorks. name for the Swift, cf. DEVIL DICK under DEVIL 2, DEVILING.

Didapper, sometimes spelt **Diedapper.** A rather widespread local name for the Dabchick, making its appearance in Ray 1678 'Didapper, or Dipper, or Dobchick, or small Doucker'. It is analysable as follows: *di-* is a reduction of *dive-*, the full form present in Shakespeare (*Venus and Adonis* 86) 1593 'dive-dapper', while *-dapper* (influenced by the adjective *dapper*) stands for *-dopper,* witness *Prompt.Parv. c.*1440 'Doppar, or dydoppar, watyr byrde'.

The above records are preceded by two earlier occurrences: *c.*1290 'dufedoppene' pl., *c.*1000 'dufedoppan' pl., which consist of *dūfe-* dive- and Middle English *doppe,* Old English *doppa* diver. The term is thus tautological. The modern formations differ from the medieval on two counts: firstly, the replacement of the obsolete *dūfe-* by a related form *dȳfe-* i.e. *dive-,* and secondly, the addition of the agent suffix -ER, productive in later names. But in connection with the last mentioned innovation we note Lincs. **Divedop,** exceptionally continuing the older stage.

Dipper. Literally a bird which dips i.e. dives into water, first in 1388 'dippere *mergulus*'. A specific sense (Dabchick) is first recognizable in Ray 1678 (see under DIDAPPER); it survives locally in Yorks. In its standard sense, the present name makes its debut in Tunstall 1771 '*Cinclus* Water Ouzel or Dipper', but on what authority is unknown. Most of the early naturalists continued to use the former term (as Pennant 1768), but Selby 1825 and Fleming 1828 chose the latter and this became usual after Yarrell 1843 'Common Dipper', though his successors generally found the epithet 'Common' expendable. Cf. DIVER, DOPPER, DUCKER.

Dip Purl. A Norf. name for the Common Tern, *dip* here meaning dive, while *purl* is ultimately onomatopœic; cf. PURRE 2 and see also under TERN.

Dirten Alan, Dirty Allan. A Scottish expression for the Arctic Skua, *dirten* being a variant of standard English *dirty* in its original sense of shitty. This epithet was motivated by the bird's pursuit of other seabirds until they disgorge, the disgorged food once thought to be excrement. The present term first appears in Pennant 1771 'Dirty Aulin'. See ALAN and cf. SCOUTIE ALLAN, also KEPSHITE.

Dishwasher, see WASHTAIL.

Dive Dapper, Divedop, see DIDAPPER.

Diver. First found as an (unspecified) bird name *c.*1500. Its present standard use as a generic term began with Edwards 1764 and was established by

Pennant 1768. Also a Ches. name for the Dabchick; cf. DUCKER. See also BLACK DIVER.

Dobchick, see DABCHICK.

Doo. A Scottish form of DOVE.

Dooker, Dopchick, see DUCKER, DABCHICK.

Dopper. An obsolete name for the Dabchick, recorded between c.1440 and 1634; the literal meaning is dipper; see DIDAPPER.

Dorbie. A Scottish name for a Dunlin or some other small wader. It is formed with the -IE suffix from the dialect verb *dorb* to grub, peck, the literal sense being therefore grubber, pecker, as in the synonym PICKEREL.

Dor Hawk. A rather local southern term for the Nightjar, known since Charleston 1668, reference being to the hawk-like appearance and the dor beetles commonly taken by this bird. Cf. NIGHT HAWK.

Dotterel. This name, almost the only one known to have been used for the species in question, makes its debut in *Prompt.Parv. c.*1440 'dotrelle' defined as 1) 'dotarde', 2) 'byrde'. It has hitherto been assumed that the bird name was simply an application of the sense dotard, fool. The Dotterel makes good eating and through its tameness falls an easy victim to the fowler. This tameness was interpreted as foolishness, hence the apparently opprobrious name. We notice, however, the Norfolk synonym **Dot Plover** and recognize in *dot* the imitative representation of the distinctive note often heard at the end of the soft, fluting call. It is evident that Dotterel is of similar origin, being the expressive *dot(t)* extended by the suffix -EREL. The derogatory meaning came afterwards, arising through the chance similarity with dotard, which is, of course, a derivative of *dote*, a word having nothing to do with the echoic

root we now recognize in the bird name. For another name, see WIND.

Doucker, see DOOKER.

Dove. The name goes back regularly via Middle English *douve* to Old English **dūfe*, not actually found in the surviving records, but confirmed by the corresponding Dutch *duif*, German *Taube*, further Old Norse *dūfa*, Gothic *-dubo*. All these forms presuppose Proto-Germanic **dūvō*, root **dūv-*, imitative of the cooing of *Columba*, or more precisely of the Wood Pigeon and Stock Dove, the Rock Dove hardly coming into the ken of those who spoke Proto-Germanic.

The surprising absence from Old English sources of this otherwise common word must be essentially due to the religious character of so much of the contemporary literature. According to traditional Germanic belief, the Dove was a bird of death, and this heathen association seems to have made the very name unpalatable to the Anglo-Saxon Church, which preferred an entirely different term; see CULVER. It was not until c.1200 that the present name is first found, more and more replacing Culver in religious texts too. By about 1300 it is seen to include *Streptopelia* in the compound TURTLE DOVE. Since the 16th cent., however, the present word has lost ground to a relative newcomer, PIGEON. It is likely that the rehabilitation of Dove was assisted by Old Norse *dūfa* from the Danelaw.

Dovekie. A Scottish seaman's name for the Black Guillemot, the record going back to 1821. It is a pet form of DOVE formed with the diminutive suffix -KIE. The name was motivated by the dove-like devotion of the sexes; cf. SEA TURTLE.

Dowp. A north country name for a Crow. Origin unknown.

Drake 1. The male of the Duck, not recorded until c.1300, but indirectly attested since 1195 (see SHELDRAKE). The German evidence shows that the word is of West Germanic age and that the Old English form will have been *draca*. The relevant cognates are Low German *Drake*, High German dialect (Oberhessen) *Antracht*, properly *Anttracht*, going back to (normalized) Old High German *anuttrahho*, where the first element *anut-* means Duck, the second *-trahho* presupposing a West Germanic stem **drakan-*, which would regularly yield the Old English form posited above.

The need for a special term to denote the male would arise as a consequence of domestication. The present word will therefore be a farmyard term in origin, referring to some distinguishing feature of the tame male. This is evidently the voice, for the root **drak-* is at once comprehensible as an echoic representation of the soft, rather grating call of the male, markedly different from the loud quacks of the female—the same vocal difference is found in the Mallard, the ancestral wild form. Cf. DRAKE 2.

Drake 2. A term denoting the Corn Crake, widespread in local use. It is of imitative origin and as such closely comparable to CRAKE 1, with which it may form a doublet, as in Yorks. CORN DRAKE. There is a further Yorks. synonym **Drake Hen**, with a variant **Draker Hen**, apparently under the influence of DAKER HEN. It goes without saying that the present name could easily be associated with DRAKE 1 male duck, and such an association is evident in two other northern names for this bird: **Grass Drake**, which in turn must have inspired **Gorse Duck** (*gorse* being here a dialect form of *grass*).

Drush. The Thrush in Dorset and Somerset, reflecting dialect pronunciation.

Duck. A term which did not become universal until the 15th cent., when it finally replaced a much older name ENDE. Commonly occurring Middle English forms are *doke*, *duke*, preceded by Old English *dūce*, attested only once *a*.967. The word literally means diver; cf. Old English *dūcan* to dive, whence our modern verb *duck*. There can hardly be any doubt that Duck established itself as the standard expression for the simple reason that the old name came to sound exactly like end.

As a generic term, Duck enters into the names of certain species, whether of popular origin as SHELDUCK or ornithologists' creations as LONG-TAILED DUCK.

Ducker. Essentially a general name for any diving or ducking bird, but most commonly attached to the Dabchick. It is first found *c*.1475 '*Mergulus* a dokare', also still current locally, especially in Scotland, in the archaic form **Dooker**, often spelt **Doucker**; see further under DUCK. Cf. DIVER.

Dulwilly, that is to say **Dull Willy.** A fanciful East Anglian name for the Ringed Plover, suggested by the musical call—unadulterated onomatopoeia, on the other hand, is found in the synonymous TULLET.

Dun, also written **Dunne.** A Ches. and Northern Ireland (Belfast Lough) name for the Knot, from the dun or dull brown colour of its summer plumage. Such use of an adjective as a noun implies a medieval origin, cf. REE (under REEVE), SWIFT.

Dung Bird. A name for the Hoopoe from Charleton 1668, motivated by the filthy nest. Also a Skua name, it being once believed that the disgorged food seen falling from the birds victimized was actually excrement; cf. TOD BIRD.

Dunlin, formerly **Dunling.** The earliest records are 1530–4 'Dunling' and 1531–2 'Dunlin', also in Ray 1678 'North-Country Dunlin', whence Pennant 1768 'Dunlin', since which time it may be regarded as the standard term. So spelt, the name reflects dialect pronunciation; it has been formed with the suffix -LIN for -LING from the adjective *dun*, literally meaning dun bird. Cf. DUNNOCK.

The bird in question shows marked seasonal changes in plumage and often in habitat, so that early naturalists thought they were dealing with two separate species. Linnaeus 1758 distinguished between *Tringa alpina* (the dun bird on the high moors in summer) and *T. cinclus* (the greyish-white subject by the shore in winter). Pennant's 'Dunlin' above appropriately refers to the former, the latter he calls 'Purre'. This remained common practice until Temminck in 1815 united the two as *T. variabilis*. After this, the English terms were often quoted as alternatives, e.g. in Selby 1833 and Gould 1837 '*T. variabilis* Dunlin or Purre'. But Jenyns 1835 had used 'Dunlin' only and Yarrell 1843 also decided in favour of the present name which, from then on, was accepted as the standard, PURRE falling into disuse as a literary term.

Dunne, see DUN.

Dunnock. A name for the Hedge Sparrow occurring in various parts up and down the country. The earliest records are *c.*1475 'donek' and 1483 'Dunoke', later 1611 'Dunneck', 1824 'dunnock'. The word consists of the adjective *dun* and the suffix -OCK, literally meaning dun bird, as also DUNLIN. During the last twenty years, a number of writers have used the present term in preference to HEDGE SPARROW.

Dun Pickle, Dun Piddle. Names for the Marsh Harrier, the former from Wilts., the latter from Dorset, recorded in 1873 and 1863 respectively. *Pickle* and *Piddle* are corruptions of obsolete *pittle buzzard, kite; see further under PUTTOCK. The adjective *dun* naturally alludes to the colour of the plumage.

Dunter. An Orkney and Shetland term for the Eider. These diving Ducks are commonly observed just off shore, bobbing up and down in seas that look rough enough to overwhelm them. This bobbing up and down was locally calling *dunting*, a word of Norse origin, from which the present name has been formed. It was first recorded in 1774. It will have replaced earlier Norn *aid (or a similar form), the traditional Scandinavian name; see EIDER.

Dun Wagtail. A Sussex name for the Grey Wagtail.

Dykie. A Scottish name for the Hedge Sparrow, formed with the -IE suffix from *dyke* i.e. *dike*, locally meaning hedge. Cf. HEDGY.

E

Eagle. Goes back to Middle English *ēgle*, known since *c*.1385; the present spelling is first noticed in 1555 'Erens or eagles'. The word came in with the Normans, cf. Old French *egle*, also *aigle*, from Latin *aquila*; it has replaced native English ERNE. See GOLDEN EAGLE.

Ear Bird. A local Sussex development of WHEATEAR, which itself arose in that county.

Eared Grebe. A term for the Black-necked Grebe coined by Pennant 1768, influenced by Edwards 1747 'Eared Dobchick'. Though replaced in standard use since Hartert 1912, the present term is still sometimes met with.

Eared Owl. Abbreviated from SHORT- and LONG-EARED OWL coined by Pennant 1768, the concept deriving from ornithological Latin *auritus* eared, itself suggested by Greek *ōtós* eared owl. For a traditional term, see HORNED OWL under HORN OWL.

Earl Duck, see HARLE.

Ear Tick. A Somerset name for the Whinchat, *tick* being imitative of the call note, as in UTICK. The epithet *ear* finds its parallel in EAR BIRD, a term for the Wheatear, the crossing of names being evidently due to the often observed confusion between *Oenanthe* and species of *Saxicola*, discussed under STONE CHACK.

Easing Sparrow, Eaves Sparrow. Local names, from Shrops. and Notts. respectively, for the House Sparrow, which often nests under the eaves; see further under EASING SWALLOW.

Easing Swallow, Eaves Swallow. North country names for the House Martin, self-evident designations for a swallow-like species which so frequently builds under the eaves. The word *easing*, common in the local speech of the northern part of the country, is a contraction of obsolete **eavesing* (1460 'evesynge'), i.e. *eaves* with secondary suffix *-ing*, like e.g. *rail* and *railing*.

Ebb Sleeper. A Shetland name for the Dunlin, *ebb* having here the local meaning of 'foreshore', a reflection of Norn idiom; cf. Old Norse *fjara* ebb-tide or foreshore.

Eckle, Eckwall, see HICKWALL.

Eggister, see HAGGISTER.

Egret. First attested in 1411 'Egrets'. Pennant 1768 used the term to denote 'a species of Heron now scarce in this island', but in 1770 qualified it as **Little Egret**, i.e. little by comparison with the Great White Heron, formerly regarded as of the same genus and named *Egretta alba*. Pennant's revised term was taken up by Bewick 1804 and Gould 1837, becoming standard with Yarrell 1843.

The name is French, occurring in Anglo-Norman 1366 *egret*, in other early French *aigret*, *aigrette* literally little heron, arising by suffix change from *aigron* (i.e. with substitution of diminutive *-et*, *-ette* for original *-on*), a phonetic spelling of **haigron*, the form lying behind HERON, initial *h* in French being dropped from about the beginning of the 14th cent.

Eider, also Eider Duck. This species was first scientifically described by the Danish naturalist Worm 1655, his name 'Eider' being a phonetic representation (in terms of Danish spelling) of Icelandic *æður*. Ray 1678 referred to this bird as 'Wormius' his Eider, or soft-feathered Duck, St. Cuthbert's Duck'. Pennant 1768 adopted Worm's term, adding DUCK for systematic reasons, whence Eider Duck, which was taken up by virtually all his successors. It was used in BOU 1883 and remained general until the present century, when the tendency has been to shorten to Eider, Hartert 1912 having 'Common Eider'.

The Icelandic source word *æður* continues Old Norse *æðr*, an abbreviation of *æðarfugl* literally down bird, the name thus primarily denoting the female (as it still does in Iceland), for it is she who lines the nest with her incomparable down. For a similarly motivated synonym, see COLK, and for a traditional English name, CUTHBERT DUCK; cf. also DUNTER.

Fróð. xxii 112–15 (1974)

Elk. A name for the Whooper Swan, known since 1552, for long in use in the Fens and in the north country and recorded, in the last century, in Antrim and Down too. It is quoted by Ray 1678 'a wild Swan, called also an Elk, and in some places a Hooper', but avoided by Pennant 1768; see WILD SWAN.

This word came to us with the (Norwegian) Vikings. Archaic Old Norse *elft* swan regularly changed in later language to *elpt* which, in accordance with a well-attested sporadic change of *pt* to *kt*, further changed to **elkt*. Passing into our language, the consonant cluster was simplified, giving Elk.

Old Norse *elft* is related to Old English *ielfet* (a word which has left no later traces) and to Old High German *elbiz*, whence modern German dialect (Swiss) *Elbs*. This Germanic name is paralleled, approximately, in Slavonic, witness Russian *lebed'* or Slovene *labod*. Literally, the name means whiteness i.e. white bird, the use of abstract nouns to name birds being an archaic feature. Cognate with Latin *albus* white, this name will have been in existence before 1000 BC. In the Germanic languages, however, this very ancient name has been largely replaced by the descendants of Proto-Germanic **swanaz*; see SWAN.

Ember Goose. Often quoted in bird books, this name for the Great Northern Diver comes from the lost Norn language of Orkney, first attested in Sibbald 1684. It is closest to Norwegian *imbre*, the word GOOSE being a secondary accretion.

The ancestral Old Norse form occurs only in one manuscript, written 'himbrin', but certainly a scribal error for *himbrimi*, still the living form in Icelandic, originally **hím-brimi* literally surf-roarer. Diver names most commonly derive from the powerful, uncanny voice, and a formation like the present would arise as a consequence of name taboo; cf. BUNIVOCHIL.

Tradition has it that this bird appears off the Norwegian coasts during the *imbredagar* ember days, just before Christmas. The same observation must once have been current in Orkney, for the name as we have it, like its Norwegian counterpart, is seen to have been modified by the religious term.

ZAA xix 54–6 (1971)

Emmet Hunter. Known today as a Somerset term for the Wryneck (*emmet* ant), ants figuring prominently on the menu. It is probable that the name was formerly more widespread; it first occurs in Charleton 1668.

Ende. Last recorded *c*.1475 and now replaced by DUCK, this traditional name goes back to Old English *ened*, cf. Dutch *eend*, German *Ente*, further Old Norse *önd*, of pre-Germanic age as shown by the cognate Latin *anas*, stem *anat-*. See GOOSANDER.

(-er. An agent suffix, first seen in a bird name in Late Old English *fiscere* fisher; see KINGFISHER. Subsequently becoming commoner, it appears to have reached its maximum development at the beginning of modern times. Examples of standard names include DIPPER, FLYCATCHER, WOODPECKER, among the plethora of local names BILTER, CHEEPER, FLUSHER, POKER/POACHER, TEASER. The suffix may be used to make, or attempt to make, some sense of an item which had somehow become obscure, as HEATHER BLEATER from HEATHER BLEAT, or THRESHER from THRESH. It occurs as an Anglicization of the French suffix *-ier*, see PLOVER, or of

Gaelic *-(a)ir*, e.g. in SCRABER. It may replace other endings, as JENNER HEN beside JENNY WREN, or represent a certain pronunciation, as **Cadder** for CADDAW. Sometimes the suffix expresses the idea of pertaining to, thus BASSER, HEDGER, somewhat similarly SCOTER, properly ***Sooter**. It can form abbreviated names, hence BRANTER, GOOLER from BRAN TAIL, GOOL FINCH. Finally, it may be added redundantly; cf. THRUSHER from THRUSH.)

(-erel. A suffix of French origin, the first two examples in bird names being COCKEREL and DOTTEREL, both from *Prompt.Parv. c.*1440. It became moderately productive, cf. PICKEREL, and notably in Whimbrel names associated with onomatopoeic verbs ending in *-er*, thus CHICKEREL, TITTEREL, WINDEREL. Contracted forms, giving *-rel*, easily arise and can become the standard, as indeed in WHIMBREL itself, also PETREL. This distinctive ending has occasionally reshaped the early forms of other names; see KESTREL.)

Erne. The native English name for an Eagle, going back via Middle English *ern* to Old English *earn*, comparable to Dutch *arend*, formerly *aren*, and older German *Arn*, further to Old Norse *örn*. The name is of Indo-European age, its base being identified in Eagle names in Celtic, e.g. Welsh *eryr*, in Slavonic, e.g.

Russian *oryol*, and in Lithuanian *erelis*, cf. also Greek *órnis* bird. The Greek meaning will be the primitive one, the use of the word to denote an Eagle being a case of name taboo: the formidable predator was simply called Bird, the vagueness of the appellation constituting its evasive character.

The present word survived best in Scotland, but was even there yielding to EAGLE by the 16th cent., except in the specific sense of White-tailed Eagle, first noticed in Turner 1544. It occurs in Orkney and Shetland where, however, it is a relic of Norn (descending from Old Norse *örn* above) which had fortuitously evolved the same form as Modern English; in this area the name naturally refers to the White-tailed Eagle.

Ess Cock. A north-eastern Scottish name for the Dipper, *ess* being a local word for a waterfall, taken from Gaelic *eas*.

(-et. A diminutive suffix going back to Middle English *-et*, *-ot*, of French provenance. It occurs early among bird names, first *c.*1225 'piot', now PIET. It was occasionally productive, see PUFFINET, and presumably conveyed a sense of endearment or familiarity commonly found with diminutives.)

Etwall. A Ches. name for the Green Woodpecker, a phonetic variant of ECKWALL; see under HICKWALL.

F

Falcon. This name came to us in Norman times, the early attestations reflecting contemporary French: *c.* 1250 'faucun', *c.*1330 'faucoun', 1362 'Faucons', and such forms occur until 1525 'fawcons'; the Old French *faucun, faucon* developed from Late Latin *falcō*, stem *falcōn-*. In the 15th cent., in accordance with the learned fashion of the day, the *l* of the original Latin was introduced into the English spelling, hence *c.*1425 'falconez', 1450 'falcon', this practice becoming general in the course of the next century. Eventually, this purely orthographic *l* came to be sounded, though some falconers may still pronounce the word without the *l*, as everybody does in the surname Faulkner.

The word was first employed as a general term for any bird trained for falconry. However, the sense could be narrowed to designate the Falcon par excellence, i.e. the Peregrine, and traces of this usage are found in early works, as Merrett 1667 'Faulcon'. In the nature of the case, Falcon competed to some extent with the native word HAWK, but remaining on the whole more technical or bookish. In the present generic sense, it is naturally equated with ornithological *Falco*.

Latin *falcō*, not met with before *c.*300 AD, has the appearance of a neologism arising with the then novel art of falconry. As it stands, the name seems to be connected with *falx*, stem *falc-* sickle, being understood as literally meaning sickle-bearer, where sickle could plausibly be a nickname for, say, the curved talon. On the other hand, it is at least equally possible that *falcō* is not the original form of the name, but rather the product of folk-etymological adjustment to *falx*, in which case one is entirely in the dark as to its true affinities.

With the spread of falconry, the term reached the Continental Germanic languages, but hardly before the departure of the Angles and Saxons for Britain in the second half of the 5th cent., since it is unknown in Old English; see GYRFAL-CON.

Falk. A Hebridean name for the Razorbill, first in Martin 1698. It is simply local Gaelic *falc* for more original *alc*, a borrowing from Old Norse *alka*, and is thus ultimately the same word as AUK. This exotic term became known chiefly through its being quoted by Pennant 1768. On the development of *f* in Gaelic, see TULIAC.

Fallow Chat. A name for the Wheatear, found locally from Surrey to Somerset, and also recorded from Scotland (Banff). The bird commonly frequents fallow land (cf. synonymous FALLOW SMITCH etc.), while onomatopoeic CHAT occurs again in STONECHAT, originally a Wheatear name.

Fallow Smitch. Known as a Northants. name for the Wheatear. It first occurs in Ray 1678 'Wheat-ear, Fallow-smich, White-tail'; see also SMITCH. Further **Fallow Smiter**, first in Merrett 1667, expressly as a Warwicks. word; the second element was doubtless originally imitative *smite*, closely comparable to Smitch. Cf. FALLOW CHAT.

Fanner Hawk. A Sussex name for the Kestrel, with motivation as in the synonym WIND FANNER.

Faskidar. A name for the Arctic Skua, first in Martin 1703. It is simply his spelling of the bird's Scots Gaelic name *fàsgadair* literally squeezer, with motivation as in TEASER. This exotic term found its way into various books thanks mainly to Pennant 1768 who mentioned 'the Faskidar of Martin'. Otherwise the Scottish name is commonly ALAN.

Feaser. A spurious name for the Arctic Skua, being a scribal error for TEASER. It goes back to Pennant 1776, who describes it as a Yorkshire name. It was occasionally copied by subsequent authors, last appearing in Montagu 1866. There are other cases of confusion between *f* and *t* in TULIAC, TULMER.

Feather Poke. Properly the richly feather-lined oval nest of the Long-tailed Tit, this term has come to be a widespread name for the bird that builds such a remarkable nest. Cf. POKE BAG.

Felfar, Felfoot, Fellfor, see FIELDFARE.

Fell Blackbird. A West Yorks. name for the Ring Ouzel, here classed as a Blackbird, *fell* being the local word for a high hill; cf. the Northants synonym **Hill Blackbird**, further North Yorks. **Moor Blackbird**, likewise referring to the upland habitat. Cf. HEATH THROSTLE, ROCK OUZEL.

Felt, Felter, Feltifer, Feltiflier, see FIELDFARE.

Fen Titmouse, see MARSH TIT.

Fern Owl. A rather widespread name locally, both in England and Scotland, for the Nightjar. This bird, which is often likened to an Owl (cf. CHURN OWL, GOAT OWL), may in the daytime conceal itself in a fern brake. The record goes back to Ray 1674.

Fieldfare. Together with variants (see below), this term goes back to Middle English *feldefare*, known as such since *a*.1300, but *de facto* attested at the end of the Old English period, *c*.1100, miswritten 'feldeware'. The name is clearly corrupt; the explanation that the meaning is somehow field-farer is just the obvious guess—and quite as obviously improbable as a specific bird name, for dozens of species fare over fields. The task is to find what lies behind the corruption.

The term, a compound word, occurs only in English. This strongly suggests that the name arose in this country, in which case the elements which compose it may well be recorded separately.

An examination of the many names given to the Fieldfare shows that the bird's plumage frequently provides the motivation, as local BLUE BACK or GREY THRUSH; among foreign names there is West Frisian *feale lyster* literally grey thrush. The Fieldfare is a noisy species, hence the voice often gives rise to a name, as the local JACK BIRD. Since the call can be interpreted as a grunt or other animal-like sound, the bird is sometimes termed a pig as in Welsh *socen*, or goat in West Frisian *fjildbok* literally field goat. And, naturally enough, both notions may be combined, as Welsh *socen lwyd* literally grey pig. Assuming that similar concepts lie behind the corrupt medieval *feldefare*, we posit original Old English *fealu fearh* literally grey piglet. Such a term is matched by the Welsh name just quoted, and the adjective *fealu* is paralleled in Frisian *feale lyster*, while the corruption *field* recalls the alternative Frisian *fjildbok*. Other birds have acquired comparable names on account of their animal-like noises; see CAPERCAILLIE, HEATHER BLEAT.

The form Fieldfare, representing a plausible modernization of Middle English *feldefare* (above), is first noticed in the second half of the 17th cent.; it was used by Ray 1678 and taken up by Pennant 1768. In many parts of the country, however, the word evolved differently, connection with the stabilizing element *field-* being broken, resulting in a plethora of local forms: twenty-five in the *EDD*, a dozen more in the *SND*, from which we select typical examples. Three syllables of the original four of the medieval form may be retained: Midlands **Fildefare**, southwest **Veldiver** (with regular local development of *f* to *v*), Scottish **Feltifer**. This type has been particularly productive in Scotland, e.g. **Feltiflier, Flirty Fleer**, beside curtailed **Felt**, this last common also in various parts of England, in Surrey becoming **Felter**; see -ER. In the south-west there is also the Wilts. derivative **Velverd**, showing that Glos. **Velly Bird** arose from an unrecorded *Veliverd*. Among contracted, two-syllable forms, we have north and Midlands **Felfar** with an early record *c*.1634, West Yorks., Lancs., Cumb. **Fellfor**, a form that could be associated with local *fell* hill, East Anglian **Fullfor**, **Fullfit**, also **Felfoot**.

As a winter visitor, the Fieldfare may acquire names reflecting this status; see SNOW BIRD. Another instance is the evocative 'frosty feldefare', occurring first in Chaucer *c*.1381, with a second record from 1430. The epithet *frosty*, which evidently formed an integral part of the name, is the earliest example of a formation which has remained moderately productive, as e.g. in STORMY PETREL, SNOWY OWL, local SANDY LOO.

Fieldfare names are sometimes transferred to the Mistle Thrush, and vice versa.

Amster. St. (Series IV) xvi 191–4 (1981)

Finch. A traditional name which goes back to Old English *finc*, corresponding to Dutch *vink* and German *Fink*. Today loosely applied to various small birds, it originally denoted the Chaffinch, as the etymology demonstrates: the Germanic base *fink-* presupposes pre-Germanic **ping-*, a syllable at once suggestive of the montonous call so typical of the Chaffinch; see PINK, also SPINK, and the Introduction, p. 8.

Firecrest. Small numbers are known to have been breeding, notably in the New Forest, since 1961, but the species is still rare, and in former times was even rarer, so that it never acquired a folk-name. If it were noticed by a layman, he would not have distinguished it from a Goldcrest. This latter had been called by Pennant 1768 'Gold-crested Wren', while the present species was named by Temminck 1820 *Regulus ignicapillus* (*regulus* wren, *igni-* fire, *-capillus* hair). Eyton 1836 took elements from both these terms to create 'Fire-crested Wren', the development of which followed the same lines as GOLDCREST, i.e. becoming in BOU 1883 'Fire-crest' and maintaining its position as the standard

term in spite of Hartert 1912 'Fire-crested Wren'.

Fire Flirt, also **Flirt Tail**. The Redstart continually splays and shivers i.e. flirts the fiery red tail, hence the present Yorks. names.

Fire Tail. A term for the Redstart, current in several districts, both in the south and in the north, including Scotland.

Fisher, see KINGFISHER.

Flamingo. We could hardly ignore this familiar name, even though it seems that the few sightings in Britain will have related to escaped birds rather than to genuine vagrants. It must be admitted, too, that the term is exotic, and first met with in travellers' accounts of foreign parts, starting with Hakluyt 1565 'Flemengo'. The spelling in the early attestations is inconsistent and in the 17th cent. the name is commonly preceded by Latin *passer* (sometimes corruptly), e.g. 1625 'Passea, Flemincos' (*sic*), 1630 'Passer Flaminga', 1634 'Pasche Flemingo', but this was later dropped, as in Dampier 1697 'Flamingo's'.

The sources are Portuguese *flamingo*, also *flamengo*, Spanish *flamenco* literally Fleming, inhabitant of Flanders, a term applied to the bird in Spanish as early as the first half of the 14th cent. To the Spaniards of the day, the Flemish traders were the best known of the northern Europeans, their light skins and rosy cheeks contrasting with the swarthy Mediterranean type. So it came about that the bird with its white and pink received the nickname 'Fleming' (J. Corominas, *Diccionario crítico y etimológico de la lengua castellana*, 1954)

This etymology is borne out by the French name for this bird, *flamant*, first recorded in 1534. The etymologists have awkwardly connected it with *flammant* flaming, but evidently it is simply an old spelling variant of *flamand* Fleming, an identity, incidentally, voiced by Pennant 1776—before the age of scientific linguistics! The French knew the species as well as did the Spaniards, for it breeds in the south of their country, too; the Provençal term is *flamenc*.

Flap Jack. Recorded by Halliwell 1847 as a Suffolk name for the Lapwing. We postulate an earlier, unrecorded **Flap Wing* (see FLOP WING), subsequently modified to give the present, humorous name.

Fleingall. Often quoted as a Kestrel name. The word, first found in a publication of 1607, is spurious, being a printing or scribal error for STEINGALL; the shape of the letter *s* was formerly very similar to *f*.

Flirt Tail, see FIRE FLIRT.

Flirty Fleer, see FIELDFARE.

Flop Wing. Listed in Swainson 1885 as a provincial name for the Lapwing. After popular speculation had changed LAP-WINK into LAPWING, it could hardly refrain from attempting to convert the still obscure element *lap-* into something more meaningful to go with -WING. We suppose that Lapwing was first altered to **Flap Wing* (see FLAP JACK) and then further to Flop Wing under the influence of the verb *flop* move the wings heavily and loosely up and down, a sense met with locally (*EDD*) and a fair description of the Lapwing's flight. Folk etymology thus, in the end, produced not only an entirely new name out of the original material, but also a plausible one.

Flusher. A once rather widespread north country name for the Red-backed Shrike, first in Ray 1678. The literal meaning is sweeper, and alludes to the habit of flushing i.e. sweeping the tail from side to side; cf. WHITE WHISKY JOHN.

Flycatcher. A name due to Pennant 1768, a translation of ornithological Latin *muscicapa*. For traditional names, see under SPOTTED FLYCATCHER.

Foolish Guillemot. A now little used name for the Common Guillemot, the epithet motivated by the bird's relative indifference to man. Guillemots sit close together on the cliff ledges and may be slow to fly away, so being netted without great difficulty. Pennant 1768 used 'Guillemot' both specifically and generically, but rectified the inconsistency in 1770 when he introduced the present term. Subsequent writers followed suit, until Yarrell 1843 brought in a successful competitior in the shape of COMMON GUILLEMOT, but Foolish Guillemot still occurred from time to time during the rest of the 19th cent., and later.

Fool's Coat. An obsolete Norf. name for the Goldfinch, known from Browne *a.*1682, the parti-coloured plumage suggesting the jester's motley. Cf. PROUD TAILOR, SHERIFF'S MAN.

Fork-tailed Petrel. A name for Leach's Petrel, introduced by Fleming 1828 '*P(rocellaria) Bullockii* Fork-tailed Petrel'—this author disputed the propriety of naming the bird after Leach.

Fowl. The original word for bird generally, and this was its usual sense in Middle English *foul*, as always in Old English *fugol*. It is part of the Germanic heritage, cf. Dutch *vogel*, German *Vogel*, further Old Norse *fugl* (see GAREFOWL), Gothic *fugls*. These forms presuppose Proto-Germanic **fuglaz*, believed to have originally been **fluglaz* with dissimilatory loss of the first *l*, since such a form can be connected with the verbal root seen in Old English *fleogan* to fly, fowl thus plausibly having the literal sense of flier.

The present word is commonly used in the Bible of 1611, as when Goliath says to David 'I will give thy flesh to the fowls of the air' (1 Sam. 17:44). It was still common in the 18th cent., see further under BIRD, but today occurs chiefly as a collective, e.g. *waterfowl*, *wildfowl*, also in connection with domesticated species, as *guineafowl*, *peafowl*. In northern English, however, the plural fowls is often heard when elsewhere one says hens or chickens. Otherwise this ancient word is normally met with only in regional or local names, as GAREFOWL, LICH FOWL, MOOR FOWL, RAIN FOWL.

Frank. A local term for the Heron, occurring sporadically from Stirling to Suff., the bird's harsh call being interpreted as a Christian name.

French Partridge. A popular name for the Red-legged Partridge, a species introduced from France about 1770. See RED-LEGGED PARTRIDGE.

French Pie. A Shrike name, first noticed in 1677, in local use in Staffs. *French* here means exotic; as for PIE, Shrikes in other countries not infrequently take Magpie names, as standard French *pie-grièche*. Elsewhere (Leics., Surrey) the present term is used of a Spotted Woodpecker. Cf. MURDERING PIE.

French Pie Finch. A local West Midlands name for the Brambling; the epithet *French* means foreign and characterizes the migrant bird; see PIE FINCH.

French Yellowhammer. A Devon term for the Cirl Bunting reported by Yarrell 1843, *French* implying the unusual, exotic.

Fuckwind, see WIND FUCKER.

Fullfit, Fullfor, see FIELDFARE.

Fulmar. Until about 1870, this species was rare in our latitudes, only coming to land at its nesting sites on St. Kilda. Here Viking colonizers in, say, the 9th cent., got to know the bird at close quarters, calling it *fūlmār* literally foul gull, doubtless from the bird's habit of ejecting the oily contents of its stomach on the approach of the fowler. The name then passed into the local Gaelic when this language superseded Norse in St. Kilda (perhaps in the 13th cent.), whence it entered English as a book word in Martin 1698. It was taken up by

Pennant 1768 in preference to MALLEMUCK and so established itself in the mainstream of English nomenclature. Since the name of the bird was originally Norse, the Fulmar must have been nesting on St. Kilda as early as the Viking age, remaining stationary until the dramatic expansion *c.*1870.

In Gaelic the word is spelt *fulmair*, the final syllable influenced by the secondary ending *-air*, a usage paralleled in English *-ER*. Martin could (should!) have written **Fulmer**—a spelling found in Montagu 1802—with the same ending as he has, e.g. in SCRABER (Gaelic *sgràbair*), cf. also Macaulay's TULMER. See MAW, WHITE PUFFIN.

Yarrell 1843 called this bird **Fulmar Petrel**, a name which has since gained a firm place in ornithological literature; it occurs in Hartert 1912 and *Checklist* 1952. But Fulmar is adequate for ordinary purposes, and entirely explicit, as in BOU 1883.

BB xlvii 336–9 (1954)

Furze Chat. A regional equivalent of WHINCHAT, in use from Sussex to Cornwall, and similarly also applicable to the Stonechat.

Furze Chirper, also **Furze Chucker.** Reported by Halliwell 1847 as names for the Brambling, to be heard in Hants. We notice, however, a number of comparable terms, from various parts of the country, very appositely used to name the Whinchat or the Stonechat, e.g. FURZE CHUCK, FURZE CHAT; see also WHINCHAT. We conclude that the present names likewise properly refer to these common species too, and have been here incorrectly applied to the not so dissimilar Brambling, which is, indeed, a rather rare visitor to Hants.

Furze Chuck. An East Anglian term for the Whinchat or Stonechat, in Hants. **Furze Chucker** (as explained under FURZE CHIRPER). Cf. FURZE CHAT.

Furze Clacker. A Surrey name for the Stonechat, doubtless also applicable to the Whinchat, motivated by the habitat and the voice.

Furze Hacker. A Hants. name for the Stonechat and the Whinchat, *hacker* being ultimately onomatopoeic, as typically in names for these species constructed after the present pattern; it is close to FURZE CLACKER. A Sussex variant is **Furze Hackle**. See GORSE HATCH.

Furze Jack. A Sussex term for the Stonechat or the Whinchat, in Glos. corrupted to **Fuzz Jack**. The original form was *Furze Chack; cf. FURZE CHUCK, and see under STONE CHACK.

Furze Wren. A Surrey term for the Dartford Warbler, occurring in Yarrell 1843, an appropriate folk-name for a bird, wren-like in appearance, living on furze-covered heathland.

G

Gadwall. First in Ray 1674, again in 1678 'Gadwall or Gray', taken up by Pennant 1768 and subsequent writers. The variant **Gaddel**, occurring in Merrett 1667, leaves no doubt that this is basically an onomatopoeic name, not surprising in the case of this incessantly chattering species; cf. scientific *Anas strepera* literally noisy duck.

Gairfowl, see GAREFOWL.

Galley Bird. A Green Woodpecker name from Kent, Surrey, and Sussex, the epithet evidently based on Middle English *gale* noise, chatter. Several names for this species were inspired by the call, see under YAFFINGALE.

Gallinule. Introduced by Pennant 1776 as a generic term embracing the Spotted Crake, Corncrake, and Moorhen, the source being Latin *gallinula* literally chicken, a diminutive of *gallina* hen derived from *gallus* cock. It was adopted by Gesner 1555 to name the Moorhen, presumably suggested by his native Swiss German *Wasserhüenli* literally water chicken. Latham 1785 and others followed in Pennant's footsteps, more or less, so that the present term gained some currency in bird books, especially as COMMON GALLINULE. In this century, however, Gallinule has been chiefly used as a generic name for various exotic species.

Gander. A traditional word, harking back through Middle English *gandre* to Old English *gandra*, earlier *ganra*, with cognates in Dutch *gander*, the same in German (Bavarian) dialect, i.e. with excrescent *d* as developed in later Old English, but Middle Low German *ganre* representing the more archaic stage reflected today in Yiddish (Jewish German) *goner*. One may confidently postulate an ancestral West Germanic stem **ganran-*.

The origin of this stem remains a matter of speculation. As we see it, two main possibilities suggest themselves. The stem could be considered as a *-ran* elaboration of a primary root *gan-*, perhaps echoic and maybe also present in GANNET. Or it could represent a Proto-Germanic **ganzan-*, earlier **gansán-*, the consonant changes envisaged (*s>z>r*) being regular in this environment. Following the latter possibility, the word appears as an extension of that most ancient stem *gans-* known to lie behind GOOSE.

Gannard. A name for the Gannet, known from a single source dated 1556, formed from GANNET, the termination -ARD being substituted for -ET.

Gannet. A traditional word, Middle English *ganet*, Old English *ganot*, with close cognates in Dutch *gent* and Swiss German *Ganz*. These continental terms, masculine in gender (like the Old English) mean gander, but in Old English this original meaning was widened to include the female, i.e. the word came to mean goose in the general sense. There is no evidence for the meaning gannet in Old English times, and hence the often quoted Old English *ganotes bæþ*, a poetical paraphrase for the sea, could not have been literally gannet's bath, as so often stated; the meaning was goose's bath.

This general sense goose continued into the Middle English period, where it is recorded as late as 1450. Meanwhile, in connection with the (former) gannetry on Lundy Island in the Bristol Channel, the word was aquiring its present sense, the first attestation dating from 1274 in the Latinized name *Petra Ganetorum* Gannets' Rock, the actual site of the colony. Now the Gannet is a very striking bird, quite distinct from other species, and people who know it give it an equally distinctive name. But when those on Lundy, who knew the bird best, simply called it goose, then that must have been a fowler's term, evasive in the first place at least.

Has one any idea of what original name was thus tabooed? It is evident that Lundy Island was once a Viking stronghold, for the name Lundy itself represents Old Norse *Lundey* literally Puffin Island (*lundi, ey*). It is not unreasonable to suppose that Old Norse *súla* gannet (see SOLAN) was also in use and passed into the English when the Norse language ceased to be spoken there about the end of Viking times, to be eclipsed in due course by the present name.

The word in this specific sense would for long remain confined to Lundy and the south-west, the only area in England where the bird was well known. But with the dissemination of ornithological knowledge, the term gained wider currency, as evidenced in the second known attestation *a*.1490 'ganettys'. By this time, the old general sense goose will have been passing out of use. But Gannet in its new, specific sense faced competition from a Scottish name SOLAN. The early English naturalists may mention both names, but Pennant 1768 adopted Gannet, though not until Yarrell 1843 did it become the generally accepted main term. See GANT, GAUNT.

The present name had, in Germanic times, a feminine counterpart evidenced by Latin *ganta*, expressly stated by Pliny *c*.77 to have been a Germanic word denoting some sort of Goose. The prehistory of these names is, however, obscure. Whereas it is natural to assume some connection with the West Germanic stem *ganran-*, known to have been the antecedent of GANDER, in neither case is the mode of formation sufficiently clear. The etymologist can only speculate that, in the light of the comparison with *ganran-*, the forms in question here could consist of a base *gan-*, perhaps imitative, enlarged by a *t* suffix of undetermined origin and function. Cf. also GOOSE.

ZAA xxi 416–18 (1973)

Gant. A name for the Great Bustard found in 15th and 16th cent. sources: *Prompt.Parv. c*.1440 'gantte, *Bistarda'*, 1546 'gantes or bystardes'; it is a special application of Middle English *gant* goose; see further under GAUNT.

Garden Warbler. First known to British ornithology as PETTYCHAPS, the present name makes its debut in Bewick 1832. It is a rendering of Gmelin 1793 *Sylvia hortensis*. The new term was taken up by Gould 1837 and Yarrell 1843 and has since remained the accepted term, even though it was eventually recognized that Gmelin's species is, in fact, the Orphean Warbler.

Garefowl or **Gairfowl.** The traditional name for the now extinct Great Auk, the former spelling first in print in Sibbald 1684, the latter in Martin 1698; there is, however, an earlier trace of the word *c*.1549 '*Avis* Gare dicta' (*OED*). But the term is very much older still, for it stems from the language of the Norsemen, who encountered the bird when they settled in the islands off the north of Scotland at the beginning of the Viking age, i.e. towards 800. They called it, in fun as it were, *geirfugl*, which was one of their names for that matchless flier the GYRFALCON; see this name. But, given the mentality of the age, this would be no mere witticism. The Norsemen had never seen a flightless bird before. It would be a relief to be able to refer evasively to such an uncanny creature; in other words, *geirfugl* was taken as a noa word. It came down to us through Norn.

The term has been but little used in ornithological literature, though Newton 1896 clearly favoured it.

ZAA xvii 258–62 (1969), *Neuphil. Mitt.* lxxix 396 (1978)

Garganey. First recorded in 1668 misspelt 'Gargane', then Ray 1678 'Garganey', also used by Pennant 1768, the name thus establishing itself in the standard nomenclature. It is taken from Gesner 1555 who quotes 'garganey' as the name used in the Bellinzona district

(Tessin, Switzerland). Professor G. C. Lepschy informs me that this is the local diminutive form comparable to *garganello*, a North Italian name for this species. It is based on an onomatopoeic root *garg-* echoing the raucous call. Native English names include CRICK, CRICKET TEAL, SUMMER TEAL.

Gaunt. A Lincs. name for the Great Crested Grebe. It is a local development of Middle English *gant*, contracted from *ganet*, Old English *ganot* goose. Given the appearance and habitat of the species in question, it is not surprising that it should receive such a name; cf. CARR GOOSE.

The present word is etymologically identical with GANNET, and indeed has been apparently recorded in this sense in Hakluyt 1600 'A great White foule, called of some a Gaunt' (*OED*). Cf. also GANT.

Gaverhal. A Devon name for the Jack Snipe, but doubtless equally applicable, if not more so, to the Common Snipe, as the etymology shows. The word is a relic of the Celtic language once spoken in Devon, and literally meaning goat (of) marsh, cf. Cornish *gavar* goat, *hâl* moor, originally (salt) marsh. Such an expression has been motivated by the unmistakable drumming, here likened to a goat's bleating, as in HEATHER BLEAT. Seeing that Celtic in Devon has been extinct for something like a thousand years, the present name must of course be even older.

Gaw Thrush. A Northants. name for the Mistle Thrush, *gaw* being a representation of the harsh call.

Geir. Given by Turner 1544 as the English name for a Vulture, but the word is in reality simply German *Geier*. This exotic expression was taken up by a few authors and also found its way into the Bible of 1611, e.g. Lev.xi.18, as 'Gier eagle'. On German for English in Turner, see under SISKIN.

Gerfalcon, see GYRFALCON.

Gill Hooter. A name for an Owl, especially the Barn Owl, making its first appearance in Ray 1674 'A Gill-howter Chesh. an Owl', the present spelling since *c.*1746 'Gillhooter'. The name is also known from Lancs., Staffs., Shrops., and East Anglia, commonly also **Gilly Hooter.** The first word is the Christian name *Gill*; cf. MADGE as an Owl name.

Giller Wren. A Lincs. term for the Wren. We suppose that *giller* contains the Christian name *Gill* extended by the -ER suffix occurring in similarly structured Wren names, as CHITTER WREN, TITTER WREN, also JENNER HEN. Since *Jack* was used to indicate smallness, as in JACK SNIPE, it is conceivable that the female counterpart *Gill* could have had the same function.

Gilliver Wren, with a variant **Julliver Wren.** Lincs. names for the Wren, developing from GILLER WREN. The now archaic *gilliver* means wench, hussy.

Gilly Hooter, see GILL HOOTER.

Gladdie. A name from Cornwall and Devon for the Yellowhammer. It is a contraction of **Go-laddie,** an older form of the name surviving locally in Devon, originally *Gold Laddie.

Glaucous Gull. Brünnich 1764 coined the term *Larus glaucus* greyish-blue gull, whence Pennant 1785 'Glaucous Gull'.

Glede, since the late 17th cent. sometimes spelt **Glead,** thus Ray 1678 'Kite or Glead'. Apparently our oldest name for the Kite, going back via Middle English *glede* to Old English *glida*. It corresponds to Middle Low German *glede* and is thus of West Germanic age at least; a related form occurs in Old Norse *gleða*. The word literally means glider, a most apposite designation, to quote Pennant 1768 'Sometimes it will remain quite motionless for a considerable space; at others it glides through the sky, without the least apparent action of its wings.' It has latterly been best known in the north, particularly in Scotland, where beside the regular form, an unexplained variant **Gled**, recorded since *c*.1420, has become general.

There is a tautological Ches. name **Glead Hawk**, with which compare GLEG HAWK.

Gleg Hawk. An obsolete Renfrew name for the Sparrowhawk, *gleg* meaning sharp-eyed, but evidently an alteration of original **Gled Hawk* properly the Kite, as Ches. **Glead Hawk**, the confusion being induced, as likely as not, by the saying 'as gleg as a gled'.

Glossy Ibis. This name is first found in Pennant 1812; it won immediate acceptance, replacing **Black Ibis** going back to Ray 1678. Both terms are naturalists' creations, *Ibis* being a Greek word, in origin Egyptian. The bird actually had, however, an English folk name **Black Curlew**, in use in Norf.—it was formerly a commoner visitor than it is today.

Gnat, see KNAT.

Goat Owl. A name from Glos. and Somerset for the Nightjar. This bird is often popularly regarded as an Owl, cf. CHURN OWL, FERN OWL, while the epithet *goat* is due to contamination from the book name GOAT SUCKER.

Goat Sucker. A name of learned origin, first in Cotgrave 1611 '*Grand merle*, a Goat-sucker', translating Latin *caprimulgus* (*capri-* goat, *-mulgus* milker, sucker). Perhaps Cotgrave coined the name, earlier dictionaries paraphrasing, e.g. Cooper 1565 '*Caprimulgi* Birdes like gulles that in the night sucke goats'. The Latin term itself is a translation of Greek *aigothḗlas* (*aigo-* goat, *-thḗlas* sucker), a ridiculous name which must have been due to popular etymology.

The name was used by Merrett 1667, reported in Ray 1678 'Fern-Owl, or Churn-Owl, or Goat-sucker', and adopted by Pennant 1768 and by many of his successors, not unnaturally as it is in line with the scientific *Caprimulgus*. But NIGHTJAR was often quoted, and Yarrell 1843 chose this name, after

which the present term progressively fell into disuse.

Godwit. It is not surprising that the two British species, the Black-tailed and the Bar-tailed, so closely related and so much alike, should go under a single popular name, attestation since Turner 1544 'godwittam' (Latinized), next 1552 'godwitte', 1591 'godwit'. The name, from Lincs. according to Merrett 1667, is an example of onomatopoeia modified by folk etymology. There are variants of the same ilk: **Godwipe** and **Godwike** recorded in 1579 and 1612 respectively, now obsolete, further in Cleveland **Godwin**. We suppose that these names (and those below) arose in connection with the Black-tailed Godwit, for the Bar-tailed, though a common visitor or bird of passage, is not known to have bred here, whereas the former was once an abundant resident. It is at their nesting sites that Godwits are particularly clamorous.

Original onomatopoeia, likewise corrupted by folk etymology, appears also in **Yarwhelp**, allied to **Yarwip** i.e. **Yarwhip**, local names from East Anglia, known since 1577 and Ray 1678 respectively. Godwits have a complex vocal range; cf. BARKER, SHRIEKER.

Go-Laddie, see GLADDIE.

Goldcrest. After Ray 1678 had called this bird 'The golden-crown'd Wren', Pennant 1768 coined 'Golden-crested Wren'. Some later naturalists, apparently unhappy with Wren for systematic reasons, adopted the scientific *Regulus* as an English word, or translated it, or did both, as Yarrell 1843 'Little Golden-crested Regulus, or Kinglet'. Meanwhile Pennant's term lived on, but tended to be shortened to Goldcrest, the form adopted by BOU 1883. Hartert 1912 brought back the long-winded 'Golden-crested Wren', but the present abbreviation was in practice so well established that little could be done to affect its position as the *de facto* standard name.

Folk names include GOLDEN CUTTY,

HERRING SPINK, TOT O'ER SEAS, WOOD-
COCK PILOT.

Golden Amber. A Yellowhammer name
from Shrops. and Ches., *amber* being
originally *ammer*; see YELLOWHAMMER.

Golden Cutty. Literally Golden Wren, a
Hants. name for the Goldcrest.

Golden Eagle. A term due to Ray 1678
and adopted by succeeding authors. It
is a translation of Greek *chrysáetos*,
chosen by Aldrovandi 1603 to designate
this bird. The Greek word is known
from Aelian (mid-3rd cent. BC); it
appears to have been a legendary name,
and was at all events originally uncon-
nected with the species it has come to
denote. The significance of *chrys-* golden is unknown.

Before Ray, this bird was simply
EAGLE, thus Turner 1544, contrasting
with ERNE 'White-tailed Eagle'.

Goldeneye. First noticed in Ray 1678
'Golden-eye . . . The *Irides* of the Eyes
are of a lovely yellow or gold-colour'
and apparently coined by him. Pennant
1768 accepted the term which thus
became part of the standard nomenclature. Among popular names are CURRE
and RATTLE WINGS.

Golden Oriole, see ORIOLE.

Golden Plover. After Pennant 1768, in
accordance with precedents, had called
this bird GREEN PLOVER, he subsequently (1776) introduced the present
term, which is a translation of Brisson
1760 '*Pluvialis aurea*. Pleuvier doré'. The
change is understandable, since the
former term was also commonly
applied to the Lapwing; virtually all
later authorities follow Pennant's innovation. Cf. YELLOW PLOVER.

Golden Thrush. A name used by Edwards 1751 for the Golden Oriole, as
like as not inspired by similar terms
occurring in other languages, e.g.
French *merle d'or*.

Goldfinch. Always a well-known name,
going back to Old English *goldfinc*,

motivated by the broad, golden wing-
bands, seen in their full beauty when
the bird is flying. It occurs in Turner
1544 'Gold finche' and other early
naturalists. Ray 1678 has 'Goldfinch, or
Thistle-finch', but Pennant 1768
preferred the present name, thus
confirming it as the standard. In local
use, however, it is often applied to the
Yellowhammer; in such cases the
Goldfinch goes under one or other of its
many non-standard names, as e.g. JACK
NICKER, KING HARRY, PROUD TAILOR,
RED CAP, THISTLE FINCH.

Middle English *gold* had a twofold
pronunciation, the vowel being either
short (as in Old English) or becoming
long. The modern standard *gold* derives
from the former, the latter became
goold, a form in use until the end of the
18th cent., and which may still occur in
local variants of the present name. Thus
we notice **Gool Finch**, whence **Gool
Finch**, also hypocoristic **Gooldie**,
Goolie (see -IE), further with redundant
suffix **Gooler** (see -ER), in some places
meaning Goldfinch in others naturally
Yellowhammer.

Goold Pink, Goold Spink. Local Scottish names for the Goldfinch, cf. GOWD
SPINK. For *goold*, see under GOLDFINCH.

Gooler, Gool Finch, see GOLDFINCH.

Gool French. A Devon name for the
Goldfinch, evidently a corruption of
GOOL FINCH. Perhaps the ornate plum-
age suggested the exotic French.

Goolie, see GOLDFINCH.

Goosander, formerly **Gossander.** First
in 1622 as 'Gossander' current until Ray
1674, but superseded by Ray 1678
'Goosander', similarly Penant 1768.

In pre-scientific use, the name doubtless referred to both the Goosander and
the (Red-breasted) Merganser. An investigation into origins, however, faces
difficulties so far insurmountable. The
form Goosander is, of course, to all
appearances due to secondary association with GOOSE, while the stressing of
the middle syllable indicates corruption

from a bookish source, probably Latin *merganser*, but the primary form Gossander remains essentially intractable. Typological considerations suggest that the word could well consist of a suffix -ER added to more original *Gossand*. Beyond that one can only speculate that -*and* may represent obsolete ENDE duck or its Scandinavian equivalent *önd*, stem *and-*, *goss-* then being an (unidentified) epithet.

Goose. Continuing Middle English *goos*, Old English *gōs*, this word is etymologically identical with Dutch *gans*, German *Gans*, also Old Norse *gās*, all regularly deriving from Proto-Germanic **gans-*. With these may be closely compared Greek *khḗn*, dialect *khā́n*, presupposing Proto-Greek **khans*, further (originally dialect) Latin *ānser* for older **hanser*, Lithuanian *žąsis* indicating earlier **žansis*—the Slavonic cognate, as in Russian *gus'*, which begins with *g* instead of the expected *z*, is believed to have been influenced by Germanic. Compare also, sometimes with change of meaning, Sanskrit *haṁsá-* goose, duck, swan, and Old Irish *géis* swan reflecting a Proto-Celtic stem **gansi-*. The basic form lying behind the above is Indo-European **ghans*; such a word must have come into existence by 3000 BC, at the latest, making Goose arguably the most ancient bird name in our vocabulary. See GANDER, GANNET.

Gor Cock. A northern and Scottish name for the male of the Red Grouse, known since 1620. The name is possibly an imitation of the barking call, from which the female **Gor Hen** was later abstracted.

Gore Crow. An Oxon. name, known since 1605 'gorcrow', the local equivalent of CARRION CROW, the word *gore* here retaining an otherwise obsolete sense of carrion, going back to Old English *gōr*. Cf. KET CROW.

Gor Maw. A name for the Cormorant, best known in Scotland, but reported from Northumb., Cumb., and Yorks. as

well. We regard the form as a popular perversion of the literary name CORMORANT, the element *gor* occurring e.g. in GOR COCK, while MAW is, of course, a northern term for a Gull. That the latter is an incongruous name for a Cormorant underlines the folk-etymological character of the corruption.

Gorse Chat. A Yorks. and Westmor. equivalent of WHINCHAT, and likewise also applicable to the Stonechat.

Gorse Duck, see DRAKE 2.

Gorse Hatch. Given as a Yorks. name for the Whinchat, but doubtless also used of the so similar, and in general more prominent, Stonechat, as is usually the case with popular names for these species. Comparison with synonymous FURZE HACKER indicates that *hatch* is ultimately onomatopoeic.

Goshawk. A traditional name going back via Middle English *goshauk* to Old English *gōshafoc* literally goose hawk. Apparently a falconer's term in the first place, implying that the bird was flown against Geese. This has, however, sometimes been thought improbable, and in many parts of the country the present name refers to the Peregrine Falcon, a species perhaps more likely to attack Geese. In this connection, compare Old High German *ganshabuh* denoting a species of Falcon, almost certainly the Peregrine. Although the naturalists have always used Goshawk in its present specific sense, it seems probable that it was, in fact, originally a name for the Peregrine. Old Norse *gāshaukr* (exact species undetermined) appears to have been modelled on Old English in the Viking age; see SPARROWHAWK.

Gossander, see GOOSANDER.

Gourder. A Kerry expression for the Stormy Petrel, in a source of 1756; it represents (apparently otherwise unrecorded) Gaelic *guadaire* literally whirler. The name has been motivated by the flight display, when these birds whirl round at high speed above their nesting burrows.

Gowd Spink. A northern name for the Goldfinch or Yellowhammer, *gowd* being the regional pronunciation of *gold*; cf. GOOLD SPINK.

Gowk. A nothern and especially Scottish name for the Cuckoo, first occurring as a nickname in 1198 'Geruasius Gokh', and again in 1202 'Simon Gouc' (Reaney 1976). The earliest direct reference to the bird itself dates from *a*.1300 'goke', later records including *c*.1450 'The goke answerd hur and sayd V times, Cukkow!' (*MED*), Turner 1544 '*cuculus* cukkouu and gowke'.

The word came in with the Vikings, cf. Old Norse *gaukr*, cognate with Old English *gēac* and Old High German *gouh*, all presupposing Proto-Germanic **gaukaz*, the onomatopoeic root *gauk* representing one note of the Cuckoo's call. A comparable one-syllable root is seen in the Celtic languages, as Welsh *cog*, Irish Gaelic *cuach*; see further under CUCKOO.

Gowk's Fool. A northern term for the Meadow Pipit, the bird being a favourite host of the GOWK, i.e. the Cuckoo.

Goylir. A Hebridean word for the Stormy Petrel from Martin 1703; it is Gaelic *goillir*, formed from *goill* blubber with the suffix *-ir* corresponding to English *-er*—the name could have been Anglicized **Goyler* (see further under FULMAR). Such a name must have originally referred to the corpulent nestling, as explained under MARTIN OIL.

Grass Drake, see DRAKE 2.

Grasshopper Warbler. In Ray 1678 designated 'The Titlark that sings like a Grasshopper, *Locustella*', which Pennant 1768 enlarged as 'Grasshopper Lark', becoming in Pennant 1776 'Grasshopper Lark Warbler', i.e. with the addition of the generic WARBLER created in 1773. This lengthy term Latham 1783 abbreviated to 'Grasshop-

per Warbler', which was accepted by succeeding naturalists. The Latin *locustella* is literally little grasshopper (*locusta*).

(Great. In older English this adjective could be used in the sense of present-day *big* and this usage was adopted by the early naturalists to coin specific terms, often reflecting Latin *major*, as explained under GREATER.)

Great Auk. Coined by Pennant 1768 under the influence of Brisson 1760 *Alca major* (instead of Linnaeus 1758 *Alca impennis*); see further under AUK.

Great Bustard. Introduced by Pennant 1776 to distinguish formally the then resident species, previously known simply as BUSTARD, from the rare visitor, the Little Bustard.

Great Crested Grebe. Coined by Pennant 1768 in modification of Ray 1678 'Greater (*also* Great) Crested . . . Doucker'. Later authorities generally dispensed with the redundant *great(er)*. But Yarrell 1843 re-introduced *great*, using it consistently, thus confirming Pennant's term as the ornithological standard.

For folk names, see CARR GOOSE, GAUNT.

(Greater. An epithet used by the early ornithologists, translating Latin *major*. However, they may also use GREAT in the same sense, often indiscriminately in the same work.)

Greater Black-Backed Gull. Pennant 1776 called this species 'Black backed Gull'; he had observed the Lesser Black-backed, but doubted whether it differed specifically from the larger bird. Montagu 1802 coined the distinguishing 'Great Black-backed Gull', a term adopted by subsequent authors, including Yarrell 1843. But the present form was used in BOU 1883, becoming standard usage. Folk names include BAAKIE, SWAABIE; see also BLACK BACK, SADDLE BACK.

Greater Redpoll. Coined by Montagu 1802 'Greater Redpole', logically contrasting with the same author's 'Lesser Redpole', but after 1837 replaced in the literature by MEALY REDPOLL.

Great Grey Shrike. Coined by Yarrell 1843, replacing Pennant 1776 'Great Shrike'; see SHRIKE.

Great Northern Diver. A term due to Pennant 1768, a conflation, so it seems, of Brisson 1760 *grand plongeon* and Linnaeus 1758 *Colymbus septentrionalis*, though this latter denoted the Red-throated Diver—the classification of the various species presented difficulties to the early naturalists. This cumbersome term Pennant 1776 reduced to 'Northern Diver' and most authorities followed suit, until Yarrell 1843 revived the original three-barrelled name to head his account of the species, though in the text itself he refers only to the 'Northern Diver'. Nevertheless, Yarrell's head-wording was seen as the standard, witness BOU 1883 'Great Northern Diver', and so it has remained.

Popular names for this bird commonly derive from the loud, wailing voice, the terms being simply echoic, as CARARA, NAAK, or originally so, e.g. ARRAN HAWK, or else more fanciful words indicating noa expressions, as BUNIVOCHIL, EMBER GOOSE.

Great Plover. A name for the Stone Curlew, apparently invented by Bewick 1797. It was generally ignored by later writers until Yarrell 1843, etc. surprisingly made it his main term. But after BOU 1883 'Stone Curlew' nothing more was heard of the present coinage.

Great Shearwater. For long known to the Scillonians (see HACKBOLT), this bird was not recognized scientifically as a British species until a specimen, obtained at the mouth of the Tees, was exhibited in 1832. It was described by Yarrell 1843 under the name 'Greater Shearwater', a translation of *Puffinus major* coined by Faber 1822. This name

appears in BOU 1883, but was altered in Hartert 1912 to the present form.

Great Skua. This name we first note in Dixon 1894. It was adopted by Hartert 1912 and quickly became general, latterly often with the folk-name BONXIE as an alternative. In the earlier literature, this species went under the name of COMMON SKUA.

Great Snipe. First used by Pennant 1768. This species is one of our less frequent birds of passage and, being only slightly larger than the Common Snipe, which it otherwise so closely resembles, was not distinguished specifically until the age of scientific investigation.

Great Spotted Woodpecker. The name was conceived by Ray 1678 as 'Greater Spotted Woodpecker' and amended to its present form by Pennant 1768. It has remained the official term, but is sometimes abbreviated to SPOTTED WOODPECKER; see this name.

Great Tit. Generalized by Yarrell 1843, for the older **Great Titmouse** commonly used by previous writers, being first found in Turner 1544 'great titmous or great oxei'. It is a naturalist's term, reflecting the scientific *Parus major*. For folk names, see JOE BEN, OX EYE, SAW SHARPENER.

Grebe. Introduced by Pennant 1768 as a convenient generic term, taken from Brisson 1760 *grèbe*. This French word appears to have become rather well known through the trade in skins, at the time widely used for muffs and collars. Its earlier history is obscure. As indicating a Grebe, the word is first known from Belon 1555; according to Gesner (also) 1555, it was a Savoyard name for a species of Gull.

Greenfinch. Known since Palsgrave c.1532 'grene fynche'. It appears in Turner 1544 'grene finche' and Merrett 1667 'Green-finch'. Ray 1678 notes 'The Green-finch, called in the Northern parts of England the Green Linnet'. Pennant 1768 chose 'greenfinch', to all appearances already regarded as the main term, and as like as not traditional, cf. Dutch *groenvink*, German *Grünfink*. See GREEN LINNET.

Greenile. A Glos. name for the Green Woodpecker, evidently modelled on WHETILE, with substitution of *green*- for obscure *whet*-.

Greenland Dove. A seaman's name for the Black Guillemot, recorded by Ray 1678 'Greenland Dove or Sea-Turtle'. For the motivation, see SEA TURTLE and cf. DOVEKIE.

Green Linnet. A term for the Greenfinch, first in Ray 1678 who describes it as a northern name (see GREENFINCH), but it has also been recorded from a few localities in midland and southern districts.

Green Plover. A name for the Golden Plover, first noticed in 1610; it was used by early naturalists, thus Merrett 1667 '*Pluvialis vulg.* Whistling Plover, or green Plover'. This sense is also in Ray 1678 and Pennant 1768, but was afterwards dropped in favour of GOLDEN PLOVER. Otherwise Green Plover is best known as a popular term for the Lapwing, and is often used in this sense in writing.

Green Sandpiper. Coined by Pennant 1768, the epithet *green* referring to the legs. For a folk name, see MARTIN SNIPE.

Greenshank. The species is not much in evidence to the layman and appears to have been without a specific folk-name, the present term being due to Pennant 1768, after the analogy of REDSHANK.

Green Woodpecker. First seen in Merrett 1667 'Green Woodpecker, or Hickwall', it was the main term in Ray 1678 'The green Woodpecker, or Woodspite, called also the Rain-fowl, High-hoe, and Hew-hole', to be effectively standardized by Pennant 1768.

Grey Back. A northern expression for the Hooded Crow.

Grey Duck. A Fens term for the Gadwall, vaguely descriptive of this rather dowdy Duck, its vagueness suggesting a fowler's word, perhaps evasive in the first place. An early trace of the name is seen in Ray 1678 'Gadwall, or Gray'.

Grey Felt, see BLUE FELT.

Grey Goose. Though surviving only locally in Sussex, this is nevertheless the oldest English name for the Grey Lag Goose. It goes back to Old English *grǽgōs*, with correspondences in other Germanic languages, as German *Graugans*, also Old Norse *grágás*.

Grey Lag. First noticed in Ray 1713 'Gray Lag', *lag* being a now forgotten word for Goose, in origin a farmyard term arising from the call *lag-lag-lag*, which has been widely used in summoning or driving Geese; see the Introduction p. 14. The oldest English name for this bird is GREY GOOSE.

It is highly probable that this simple name **Lag** is, in fact, of great antiquity. It can hardly be a coincidence that a comparable syllable **lak* lies behind Gaelic *lacha* duck, a name which can be shown philologically to have been in existence for at least a millennium and a half.

Z.f.celt.Phil. xxxviii 183 (1981)

Grey Lag Goose. Introduced for systematic reasons by Pennant 1768, from the earlier GREY LAG, presumably without his realizing that the new term was tautological.

Grey Owl. A name coined by Ray 1678 to denote a species described as closely related to, but distinct from, the 'common brown or Ivy-Owl'. It is now recognized that in both cases the bird referred to is the Tawny Owl.

Grey Phalarope. Coined by Pennant 1776 after Edwards 1760 'grey cootfooted Tringa'; see PHALAROPE.

Grey Plover. This name makes its debut in Merrett 1667 '*Pluvialis cinerea* Grey Plover', but is doubtless much older, and traditional, cf. *a*.1549 'Plovers grey the dosen 3s.' (*OED*). Ray 1678 uses the name, and though Pennant 1768 called this bird 'Grey Sandpiper', his innovation did not seriously challenge the present term, which was generally preferred by other writers.

The Grey and Golden Plovers are liable to be confused, hence Grey Plover sometimes denotes the Golden Plover as well in local use. In this connection cf. WHISTLING PLOVER.

Grey Starling, see BLACK STARLING.

Grey Thrush. A Somerset name for the Fieldfare, motivated by the greyish-blue feathers on the lower part of the back.

Grey Wagtail. Used by Ray 1678, apparently his translation of Aldrovandi 1603 *Motacilla cinerea*. The term was adopted by succeeding writers. See DUN WAGTAIL, YELLOW WAGTAIL.

Grosbeak. A name given to a number of small birds, especially Finches, having a large, stout bill, but sometimes restricted to the 'Common Grosbeak' or Hawfinch. It makes its debut in Ray 1678 'The common Gros-beak or Hawfinch' and appears again in White 1767 'a cock grosbeak', the source being French *grosbec* literally thick bill, first found in Belon 1555. One has the impression that the English form is purely an ornithologist's word, taken direct from Belon (cf. GUILLEMOT), but naturalized to the extent that *-beak* was substituted for French *-bec*. See also BEAK.

Grouse, formerly written **Grous**, a spelling still found in Jenyns 1835. Since the singular and the plural of this name are the same, one may wonder if the final *s* was not originally a plural ending. The first attestations point to this conclusion: they are 1531 'grows and peions' (pigeons) and 1547 'grewes', plurals for which we posit the singulars **grow*, **grew*, these to all appearances coming from Old French **grue* deduced from

Late Latin *grūta*, known from Giraldus *c*.1185 *grutas* (accusative pl.), described as *gallinae campestres* field hens. The term has not survived in French; competition from homophonous *grue* crane from Late Latin **grūa* (Classical *grūs*) most likely accounting for this. Today, Grouse is normally taken as meaning Red Grouse, the species more familiar to the general public. But this is a development of the 19th cent., for the name primarily denotes the Black Grouse, cf. Ray 1678 'Heathcock or Black game or Grous'. This bird has a powerful voice, of which our source word *grūta* is an acceptable imitation.

Guckoo. A form of CUCKOO developing in the south-west (Glos. and Wilts. to Cornwall), first noticed in 1513 'gukgo' (*EDD*); it was quoted by Merrett 1667 'Cuckow or Guckoe'.

Guillemot. A name of French provenance, known since Belon 1555, who stated that the name denoted the immature bird. It was introduced into English to name the Common Guillemot by Ray 1678. Pennant 1768 took up the word, using it both specifically and generically (see BLACK GUILLEMOT) and *de facto* established it as a standard name. Presumably the exotic term was felt to stand above the various regional names, but only during the last hundred years or so has it actually been replacing these synonyms locally. The bird name, as we have it, is the same as the French personal name *Guillemot*, a pet form of *Guillaume*, which goes back to Old French (11th cent.) *Willelm*.

The commonest native name for the Guillemot is **Willock**, first attested in 1631, in Scotland also **Willick**, further Sussex **Willy**, **Will**, Scottish **Wilkie** (see -KIE), all sometimes transferred to the Razorbill, in which connection see under MURRE.

All the above names are ultimately echoic in origin, the clear, high-pitched call of the juvenile bird being reproduced as *will*, and used to denote the species as a whole. Such a base could hardly fail to be brought into association with pet forms of Old French *Willelm* William, an everyday Christian name in this country after 1066. And, of course, the same thing happened on the French side of the Channel; GUILLEMOT, too, is ultimately the product of the same onomatopoeia.

Different imitations of the juvenile cry also occur; see KIDDAW, SKIDDAW, SCOOT, SKUTTOCK, further WEERIT. For onomatopoeic names based on the call of the adult, see MARROT, MURRE.

Guldenhead. Found in Ray 1678 'Guldenhead, Bottle-nose', allegedly names for the Puffin from the Tenby district, but evidently an editorial lapse for 'gold and red Bottle-nose', which sounds more like a description than a regular name.
TPS 1974 74f. (1975)

Gull. Makes its debut in 1430. In origin a Celtic word from the Cornish coast, spread by seafaring folk who, notably in Tudor times, came in the main from the Cornish peninsula. During the 17th cent. this exotic name largely replaced the native terms COB and especially MEW. It represents late medieval Cornish **gullen* which, on passing into the English dialects of East Cornwall and Devon, was felt to be a plural, since in those dialects *en* was a common plural ending (as in *oxen*). Consequently a new singular Gull came into being. Early medieval Cornish **gullan* regularly derives from an older, recorded form *guilan*, etymologically identical with Welsh *gwylan*. In these words, initial *g* is secondary, the antecedent form being **uilan*, which came into existence about 600 (this stage reflected in the Old Irish loan *foilenn*; see TULIAC). It presupposes Old British **ēlan-*, a word formed from the onomatopoeic root **ēl*, cf. Welsh *wylo* wailing. Gull is thus literally wailer, an appropriate name for most species, particularly the Herring Gull.

Gutter Cock. A name in Cornwall for the Water Rail, *gutter* as in GUTTER SNIPE.

Gutter Snipe. A Northumbrian name for the (Common) Snipe, *gutter* here meaning brook or channel, reference being to the habitat.

Gyp. It is not surprising that such a favourite as the Starling should, in some places, go by a pet name, as the North Riding of Yorks. Gyp, Cumb. **Jebbie**, Northants. and Ches. **Jacob**, Oxon. **Joey**. One may have the impression that these names are vaguely related and perhaps, in the last analysis, variations of the same original, whatever that might have been.

Gyrfalcon (latterly also **Gyr Falcon**) or **Gerfalcon**. The use of the name in England can be traced to 'girfaucones' in a Latin document of *c*.1209. The earliest occurrences in English-language contexts are *c*.1300–30 'gerfauk, girfauk' and 1382 'ierfakoun'; *Prompt.Parv. c*.1440 has 'Gerfaucun'. From the 15th cent., forms with etymological *l* are found, as 1450–70 'Geir Falconnis'. On etymological spelling, see under FALCON.

The source word is Old French *girfaucon, gerfaucon* (with analogical *-faucon* falcon) for the more original *girfauc, gerfauc*, the second part being earlier *-falc*. The term was borrowed either directly or via a German intermediary from Old Norse *geirfalki*; the epithet *geir*, literally spear, denotes excellence and is a native Germanic word, *falki* falcon a borrowing through German of Latin *falcō*; see FALCON. The European race of this most powerful of Falcons is essentially a Scandinavian species, the coveted birds coming especially from Norway and Iceland. The Norse word itself is first found in a 12th-cent. Norwegian source, but will be considerably older. The bird had a (doubtless secondary) alternative name *geirfugl* (*fugl* bird) which reached the far north of Britain, where it underwent a remarkable change of meaning; see GAREFOWL.

The current spelling Gyrfalcon goes back to Pennant 1768, who followed the

Latin of Aldrovandi 1603 *Gyrfalco*, which in turn reflects the influence of medieval Latin *gyrofalco*, where *gyro-* circle is a speculative substitution, known since *c.*1188. The spelling Ger-falcon appears first in 1863 and became fairly common, but in this century Gyrfalcon has been the commoner form, since BOU 1883 often divided as Gyr Falcon.

ZAA xvii 258–62 (1969)

H

Hackbolt. A Scillonian term for the Great Shearwater, first in Yarrell 1843. The word is a corruption of an otherwise lost Cornish *hakbōl* literally chopping axe, a name motivated by the distinctive flight pattern; see CLEAVER, SHEARWATER. This Shearwater does not come to land hereabouts, the present name distinguishing it locally from the breeding species COCKATHODON.
BBCS xxviii 87–9 (1978)

Hacket, Hacklet. Terms for the Kittiwake, the latter first noticed in Kingsley, *Westward Ho!*, 1855 (*OED*), the former in Swann 1912. Kittiwake names are nearly always onomatopoeic, variously reproducing the insistent cries of this clamorous species, and the same may be presumed here.

Hackmal, see HICKMAL.

Hagdown. Given as a Shearwater name from the Isle of Man; affinities unknown.

Haggess. An obsolete local term for a Magpie: 1599 'Haggesses', 1655 'Pyes, or Haggisses'. It is a borrowing from (northern) French *agace*; cf. Cotgrave 1611 '*agace, agasse* A Pie, Piannet, or Magatapie'. The French, in its turn, was taken from Old High German *agaza*, formed from the root *ag-* sharp, pointed, in reference to the distinctive tail. The present name is thus not far removed from the native synonym HAGGISTER; it owes its initial *h* to contamination from this word. See CADDESS.

Haggister. A Kent word for the Magpie, the earliest record being 1584 'haggister', then Ray 1764 'hagester'. A variant *Eggister* is reported from Lincs. and Dorset. These names are isolated survivals of the native English tradition, otherwise swamped by French PIE and its brood; they presuppose Old English *hagastre, *egistre*, with which compare

Old Low German *agastra, agistra*, doubtless also unrecorded forms with initial *h*, as in Middle Low German *hegester* beside *egester*. In the Low German dialects, the name of the Magpie is often misapplied to the Jay. It is therefore not unlikely that the *h* is due to contamination by an old name for this bird, noticed also in Old English *higera*; see under JAY. The present names are forms extended by means of a well-known suffix *-astre/-istre* from a primitive name for the Magpie surviving in Old English *agu*, the literal meaning of which is pointed (tail); see HAGGESS.

Halcyon. Early attestations of this poetic name for the Kingfisher are 1390 'alceon', 1398 'Alicion', 1545 'halcions', 1592 'halcyon'. The word has been taken direct from Classical literature: Latin *halcyon, alcyon*, transliterations of Greek *halkuón, alkuón*, a fabulous bird identified by the ancients with the Kingfisher. It was said to breed on the sea during the so-called halcyon days, allegedly a period of calm lasting for seven days before, and as many after, the shortest day of the year. The name defies further analysis; it is, however, known that the form without *h* is the more original one.

Half Curlew. A name for the Whimbrel from the Yorks. and East Anglian coasts. The Whimbrel looks for all the world like a small Curlew, while the epithet *half* indicates that smallness; similarly JACK CURLEW, further LITTLE WHAUP. See also SPOWE.

Half Snipe. A Norf. name for the Jack Snipe. This species is by far the smallest of the Snipes, and the epithet *half* alludes to this: 'Its weight is inferior by half to that of the Snipe' as Pennant 1768 (who first mentions the name) puts it.

Hammer Bleat, see HEATHER BLEAT.

Handsaw, see HERNSAW under HERN-SHAW.

Hareld. Given in various lists as a Scottish or Orkney name for the Long-tailed Duck. It was used by certain early writers, but has never been a living word, being in fact spurious. It is due to Stephens 1824, and corresponds to his ornithological *Harelda*, a misspelling of Ray 1678 *Havelda*, this also erroneous, going back via Worm 1655 to Icelandic *haferla* (pronounced /havedla/) literally *haf-* sea, *-erla* wagtail, itself a folk-etymological variant of *hávella* /hauvedla/, ultimately imitative of the loud call of this noisy Duck; cf. COLDIE.

Harle. A French name quoted by Sibbald 1684 'Merganser, Harle *Bellonii*, the Goosander'. Treated by later authors as though it were an Orkney or Shetland word, it thereby gained some local currency, especially in the sense 'Redbreasted Merganser'; it changed through folk etymology to **Earl Duck,** also **Herald Duck.** The French name, first in Belon 1555 (= *Bellonii* above), can designate either of the aforementioned Sawbill Ducks. It is not indigenous French, but a borrowing from Germanic, formed from a root *har* imitative of the harsh calls of these birds.

Harn. A Norf. term for the Heron, developing from HERN by regular change; for the phonetics, see BERGANDER.

Harrier. A generic term, extracted from HEN HARRIER by Selby 1833 to correspond to the scientific name *Circus*.

Hatcher. A Norf. name for the Hedge Sparrow, properly HEDGE HATCHER.

Hawfinch. First in Ray 1674, then 1678 'The common Gros-beak, or Hawfinch'. Pennant 1768 and most of his successors preferred to use the former name, but Selby 1825 and Gould 1837 decided on the latter and their choice was confirmed by Yarrell 1843, since when the present name has been unequivocally the standard. *Haw-* is, of

course, the berry of the hawthorn which forms a notable part of the diet.

Hawk. A traditional name for any hawk-like species, including those now scientifically distinguished as Falcons or Harriers, as well as the Buzzard and the Kite. Naturalists, however, have from the beginning tended to restrict the use of the term to the genus *Accipiter*, hence the standard names GOSHAWK, SPARROWHAWK. The name goes back via Middle English *hauk*, earlier *hauec*, to Old English *hafoc*, and corresponds to Dutch *havik*, German *Habicht* (with secondary *t*; in dialect often *Habich*), further Old Norse *haukr*. The forms attested presuppose Proto-Germanic **havukaz*, analysable as root *hav-* with extension *-uk-* (found in other bird names; see -OCK) and inflexional *-az*. The root has been compared with pre-Germanic *kap-* seen in Latin *capere* to take, seize, the present name thus plausibly meaning the bird that seizes.

Locally confused with AUK; cf. ARRAN HAWK.

Hawk's Eye. In Somerset a name for the Goldcrest, in Surrey for the Willow Warbler; it is a corruption of OX EYE.

Hay Jack. A name recorded in Suffolk and Sussex, where it is loosely applied to various species which may build their nests with hay, typically the Whitethroats, the Blackcap, and the Garden Warbler. There is comparable motivation in JACK STRAW.

Hay Suck. This name for the Hedge Sparrow, found today with variations from Glos. to Devon, looks back on a very respectable ancestry, continuing as it does the oldest known English name for this species. Medieval forms are Middle English *heisoke, -soge, -sugge*, Old English *hegesugge*, of which the first element *hege* hedge develops regularly through Middle English *hei* to present-day *hay*. This word may still be remembered in certain rural districts as a word for hedge, but generally today the word is, of course, understood as meaning dried grass (in which sense it is of a quite different origin). The proper sense of *hay* in the present name has thus, in practice, become obscure.

Nor has the second element fared any better. Indeed, as comparison with related languages shows, even Old English -*sugge* sucker was already corrupt; it will originally have been **-smugge* creeper; cf. especially German *Grasmücke* whitethroat from Old High German *gras-smukka* literally grass creeper, where -*kk*- regularly corresponds to Old English -*gg*-. There is a close semantic parallel in NETTLE CREEPER, a local term especially for the Whitethroats, and we may add, to complete the analogy, that **Hedge Creeper** is attested as a name for the Hedge Sparrow.

As noted, the second element of the name came, at a very early date, to be associated with the word suck (Old English -*sugge*, cf. *sūgan*, also *sūcan* to suck). And folk etymology continued to work havoc. In Somerset and Devon they tried to make sense out of nonsense by altering the name to **Hay Sucker**; in Worcs. it was thought that **Haze Cock** sounded better. Already in an older source *a*.1475 'eysoges', we meet a form which has dropped its *h*, and perhaps this was the antecedent of another Worcs. name **Isaac**, whence also **Molly Isaacs**, in Glos. **Blue Isaac** (see BLUE SPARROW), sometimes **Ikey**, misleadingly spelt **Aichee**.

Haze Cock, see HAY SUCK.

Heath Cock. A name for the Black Grouse, first in 1590, and also noted by Ray 1674 'The common Heath cock, Black game or Grous'.

Heather Bleat. A widespread northern and Scottish term for the Snipe, first recorded in this form *c*.1700 'Heatherbleet'. A modernizing **Heather Bleater**, with secondary -ER, has been known since 1820. There is a variant **Heather Blutter** going back to 1590 'hedderbluter', further Cumb. **Hammer Bleat**. These various names are essen-

tially folk-etymological developments of a single original name seen in Old English (11th cent.) *hæferblaete* (*hæfer* billy goat, *blǽte* bleater). The expression alludes to the characteristic drumming, that resonant neighing or bleating sound heard from the air-borne male, especially during courtship. There is comparable imagery in GAVERHAL; see also HORSE GOWK.

Heath Throstle. A Yorks. name for the Ring Ouzel, here classed as a Thrush, while *heath* refers to the upland habitat, cf. FELL BLACKBIRD, ROCK OUZEL. The present name has been known since 1676.

Heckymal, see HICKMAL.

Hedge Accentor. A term for the Hedge Sparrow coined by Selby 1825 and taken up by Jenyns 1835 and Gould 1837, also by Yarrell 1843 and succeeding editions. But this half-Latin book name found little support in the latter part of the century and quickly fell into disuse. The word *Accentor* was created by Bechstein in 1803 as the scientific name for a genus including the Hedge Sparrow; it is based on Latin *cantor* singer and was intended to mean something like warbler.

Hedge Chat. A Northants. term for the Hedge Sparrow, which occasionally takes its name from its voice; see CHAT.

Hedge Creeper, see HAY SUCK.

Hedge Hatcher. A Norf. name for the Hedge Sparrow; also HATCHER.

Hedge Hawk. A name for the Sparrowhawk, alluding to its hunting habits, the bird quartering the hedgerows in search of the small birds upon which it preys.

Hedge Jug. A name for the Hedge Sparrow from an unspecified locality, *Jug* being a pet form of the Christian name Jane; see further under JUGGY WREN. Cf. BLUE JIG.

Hedge Pick, Hedge Pip. Southern names for the Hedge Sparrow, *pick* and *pip* being onomatopoeic syllables mimicking the piping call; cf. PINNOCK.

Hedger. A term for the Hedge Sparrow, locality not given. See -ER.

Hedge Sparrow. First in Palsgrave 1530 'Hedge Sparowe', next 1570 'Pinnocke, hedge sparrow'. It is the name found in Merrett 1667, quoted by Ray 1678 and used by Pennant 1768. In the first half of the last century, the naturalists tended to avoid this traditional name for systematic reasons, often employing the artificial HEDGE ACCENTOR instead. But afterwards the present name again came to the fore; it was adopted in BOU 1883 and later lists. Latterly, however, it has lost some ground to DUNNOCK.

Hedgy. A pet form of HEDGE SPARROW heard in Northumb. For a similar formation, cf. DYKIE.

Heffald. A West Yorks. term for the Green Woodpecker, from older **Heigh Hold** (*gh* acquiring the sound /f/ as in *laugh*), itself a development of HIGH HOE.

Hegrie. A term in Shetland for the Heron, from the former Norn language, the source being Old Norse *hegri*, a variant of the word seen in Old English *hrāgra*, i.e. with dissimilatory loss of the first *r*; see HERON. The final *-ie* is not so much preservation of the Old Norse ending as the influence of the relatively recent diminutive suffix -IE.

Heigh Hold, see HEFFALD.

Hempie. A name for the Hedge Sparrow from the Scottish Borders. It is a diminutive of *hemp* (see -IE) presupposing original *Hemp Bird, *Hemp Sparrow* or the like. In the winter months, this species feeds chiefly on seeds, possibly including hemp seeds.

Hen. A constituent of several bird names, the word goes back to Old English *henn*, of West Germanic age, cf. Dutch *hen*, German *Henne*; it is the feminine of Old English *hana* cock, cf. Dutch *haan*, German *Hahn*, the literal meaning of which is crower, the Germanic root *han-* deriving from the Indo-European *can-*, as seen in Latin *canere* to crow, sound, sing. The feminine form could, of course, only arise after the literal meaning of the masculine had been forgotten. Cf. SEA HEN.

Hen Harrier, formerly **Hen Harrower.** This name makes its earliest appearance in Turner 1544 'Hen harroer'. The current form is first seen in Ray 1678 as a term for the male bird only, but in Ray 1691 used specifically 'a sort of puttock called a hen-harrier from chasing, preying upon and destroying poultry'. Pennant 1768 followed Ray 1691 and so the word in this meaning passed into the standard nomenclature.

The forms Harrower and Harrier are, in origin, purely variants of the same word, the basic sense of both being the bird which pursues and destroys.

See BLUE HAWK, RING TAIL.

Herald Duck, see HARLE.

Herl. A Scottish variant of HERN, known since *c*.1500 'herle'. The reason for the curious change of *n* to *l* is unknown.

Hern. A completely Anglicized form of HERON, known since *c*.1425. It is still widespread locally, but is now unusual in the literary language, the last notable example being Tennyson's 'I come from haunts of coot and hern' (*The Brook*, 1855). Cf. HARN, HERL.

Hernshaw, Heronshaw. Local names for the Heron, the type being first seen in Palsgrave 1530 'hernshaw'. They are corruptions of an earlier type surviving in Yorks. **Hernsew, Heronsew**, recorded since 1381–2 'heronsewes', a loan from French *heronceau* 'young heron' (*-ceau* being a diminutive suffix). A further distortion is seen in Suff. **Hernsaw**, doubtless locally also *Harnsaw** (cf. HARN from HERN), whence **Handsaw** known from Shakespeare (*Hamlet*) *c*.1602 'I know a Hawke from a Handsaw'. Presumably the English sense was at one time 'young heron', as in French, but for this there is no evidence in the records as we know them. See HERON.

Heron. The first attestations, c.1300 'hayroun' and 1302 'heyruns', are typical of the early occurrences, the present spelling making its debut in 1381, but not becoming general until the 17th cent. It long competed with its variant HERN and the derivatives **Heronshaw**, HERN-SHAW, cf. Merrett 1667 'Heron, or Hern, Hernshaw', Ray 1678 'Heron or Heronshaw', but Pennant 1768 preferred the present form and his example finalized matters.

The name comes from France, witness such medieval French forms as *heron, heiron, hairon,* older **haigron,* of Germanic origin; cf. Old High German *heigaro.* This last is evidently from **hreigaro,* i.e. with dissimilatory loss of the first *r,* and thus etymologically identical with Old English *hrāgra,* the native term replaced by the present borrowing. The Old English and German forms presuppose Proto-Germanic **hraigran-* (*g* being fricative, as in Old English and the oldest German) from earlier **hraihran-,* which in turn implies a pre-Germanic root **kraikr-* (or **kroikr-*), an onomatopoeic representation of the Heron's harsh cry. On the consonant changes, see the Introduction, pp. 7–8. Cf. EGRET, HEGRIE.

Herring Gull. A term first encountered in Ray 1678. It was taken up by Pennant 1768, who *de facto* established it as the standard. Ray's source is unknown. This Gull has no special connection with herring, though it often goes fishing—but the same is true of the Lesser Black-backed. It would appear, then, that the present name is somewhat arbitrary.

Herring Spink. Literally 'herring finch', a Suff. seamen's name for the (migrating) Goldcrest, flocks of which often alighted on the rigging of the herring boats out in the North Sea. **Tot o'er Seas**, another Suff. name for this bird, would arise in the same milieu.

Hewhole. A Northumb. term for the Green Woodpecker, going back to Turner 1562 'huehol'. Hardly an original form, but a folk-etymological offshoot of the name lying behind HIGH HOE.

Hiblin. An Orkney name for the Cormorant handed down through Norn, the source being Old Norse **hypplingr* (best preserved today in Faroese *hyplingur*), apparently an abbreviation of **hvīthypplingr* literally white **hipling* i.e. white-hipped bird. The motivation and Old Norse forms are paralleled exactly in LORIN. Cf. also HUPLIN. See -LING.

Hickmal, also **Hickmall.** Variant names for the Blue Tit occurring in Cornwall, Devon, and West Somerset. To all appearances *hick-* is a reminiscence of a verb **hick* peck, otherwise unrecorded in English, but deducible from Middle Low German *hicken* with this meaning, the motivation being the repeated pecking characteristic of this bird. It seems most likely that the original form of this name is preserved in Old English *hicemāse* (i.e. compounded with *māse* tit; see TITMOUSE), the elements *-mal, -mall* in the name today being then (unexplained) corruptions. Further corruption is seen in the variant **Heckymal**, while in yet another variant **Hackmal** one sees that *hack-* has been substituted for *hick-* which was no longer understood.

Hickwall. A name for the Green Woodpecker, first recorded c.1450 'hykwale'. The present form is often quoted by the earlier naturalists, apparently following Merrett 1667 'Green Woodpecker, or Hickwall', but Pennant 1768 chose the systematically more convenient GREEN WOODPECKER. As initial *h* was widely discarded in our dialects, the name could become **Ickwall**, whence Northants. **Ickle**, then Shrops. **Eckwall**, whence Glos. and Worcs. **Eckle**, see also ETWALL, further YOCKLE, YUCKLE. In the absence of early forms, it is difficult to contemplate any etymology, though it is tempting to see in *hick-* a trace of a lost verb **hick* peck (see further under HICKMAL) while *-wall* recalls another Woodpecker name WOODWALL; see also HIGH HOE.

High Hoe. A north country name for the Green Woodpecker; cf. Ray 1678 'The green Woodpecker, or Woodspite, called also the Rain-fowl, High-hoe, and Hew-hole', spelt in Ray 1674 'Heyhoe', in Cotgrave 1611 'Heighaw, Highaw'. Such forms are reminiscent of onomatopoeia and therefore possibly represent the call, as is often the case with Woodpecker names; see under YAFFINGALE.

Palsgrave c.1532 'the high-wale *lespec'* and the 15th cent. 'Hygh-whele *picus'* suggest a connection with HICKWALL. See HEFFALD.

Hill Blackbird, see FELL BLACKBIRD.

Hill Chack. An Orkney name for the Ring Ouzel, *hill* referring to the usual habitat, *chack* imitative of the harsh alarm call. Contrast RING WHISTLE.

Hill Lintie. A name for the Twite from Caithness and Orkney, literally Hill Linnet, in reference to the habitat. Cf. MOUNTAIN LINNET.

Hobby. Makes its debut in *Prompt.Parv.* c.1440 'Hoby', the present spelling first in 1588, but already implied in Turner 1544 Latinized 'hobbia'. Evidently the only term specifically used of this bird, it is found in the works of all the early naturalists. It has been taken from Old French *hobe*, known since the end of the 13th cent., a word associated with the verb *hober* to jump about, a reduced form of *hobeler* with the same meaning, itself borrowed from Dutch *hobbelen*. The name, which presumably arose in connection with hawking, refers to the well-known agility of the Hobby in pursuit of its prey.

Hole Dove. A term for the Stock Dove, locality not given. Unlike other Doves,

this species nests in a hole in a tree-trunk.

Holland Hawk, see ARRAN HAWK.

Hollin Cock. A Yorks. name for the Mistle Thrush, preserving in *hollin* the earlier form of *holly*, and so called from its feeding on holly berries. Cf. from the other end of the country HOLM COCK, with *holm* as a variant of *hollin*.

Holm Cock, Holm Screech, Holm Thrush. Names in the south-west (Dorset to Cornwall) for the Mistle Thrush, from its partiality to holm (holly) berries. See SCREECH and cf. HOLLIN COCK.

Honey Buzzard. A name coined by Willughby, who had found combs of wasps in the nest of this bird, appearing in Ray 1678. It was adopted by Pennant 1768 and became the standard term. Such a rare species would never acquire a folk-name.

Hood Awl. A Green Woodpecker name from Cornwall. It is a folk-etymological development comparable to obsolete c.1650 'woodhall', a variant of WOODWALL. Such forms have naturally been the starting point for another local name **Awl Bird**.

Hooded Crow. In origin a Scottish term, known since the early 16th cent., the black head and throat being the hood which contrasts with the grey of the back and underparts. The term is one of several which came into being to distinguish this bird from the Carrion Crow; see ROYSTON CROW. Pennant 1776 adopted the present name which he took from Sibbald 1684, thus *de facto* conferring upon it standard status. **Hoodie** is a commonly heard pet form.

Hoolet, see HOWLET.

Hoop 1. A name for the Bullfinch, found locally from Warwicks. to Cornwall, in the far south-west occurring side by side with **Hope,** of which it is presumably a secondary variant due to popular etymology. The word first comes into our ken in Merrett 1667 'Bull finch, Hoop, Bul Spink, Nope'.

Hoop 2. An obsolete name for the Hoopoe, sometimes transferred to the Lapwing on account of the remarkable crest common to both. It makes its debut in 1481 'The huppe or lapwynche is a byrd crested', subsequent attestations including 1601 'The Houpe or *Upupa* . . . is a nasty and filthy bird' (see DUNG BIRD), 1607 'Oopes', 1666 'Whoope', Holme 1688 'A *Upupa* . . . is in our country speech called a Whoopoo, or Whopee, or Hoopoe, and Howpe', and the last record 1708 'houp'. The source is Old French *huppe* ultimately from Latin *upupa*, echoing the call note heard as *up-up*. Charleton 1668 quotes the name as **Hoopoop**.

Hoopoe. The name is first found in Merrett 1667 'Hoopee', next Ray 1678 'Hoop, or Hoopoe', in Holme 1688 'Whoopoo, or Whopee, or Hoopoe, and Howpe'. Pennant 1768 chose the present form which then became the standard. Though clearly belonging to the same onomatopoeic tradition as HOOP 2, also **Hoopoop**, the form Hoopoe (and its variants) have not been fully accounted for.

Hope, see HOOP 1, NOPE.

Horned Grebe. A term for the Slavonian Grebe coined by Pennant 1785, influenced by Edwards 1750 'Horned Dobchick'. It was used by several succeeding authors, but Yarrell 1843 favoured 'Sclavonian Grebe', the forerunner of the more usual name found in bird books today. Nevertheless the present, by no means inappropriate, term continues to find occasional adherents.

Horner. A Yorks. (Holderness) name for the Goosander. We suppose it to have been motivated by the harsh, guttural call, here compared to the sound of a (fog?) horn. See -ER.

Horn Owl. A term for an Eared Owl, first appearing in Turner 1544 'hornoul', traditionally also **Horned Owl**, known since 1390, and which continued, essentially in literary use, well into this century. This name type survives best in Scotland; see HORNY HOWLET.

Horn Pie. An East Anglian name for the Lapwing, *horn* alluding to the crest, PIE Magpie, in reference to the pied appearance. Cf. HORNY WICK.

Horny Howlet, also **Owlet, Owl.** Scottish names for an Eared Owl. See HORN OWL.

Horny Wick, Horny Wig, Horny Wink. Names in Cornwall for the Lapwing. The substitute *horny* for original *lap-*, a reference to the horn-like crest, is due to contact with Celtic Cornish CORNWILLEN, see further under this name. The varying second elements illustrate the continual working of folk etymology. Of these, Horny Wink preserves the oldest form reflecting obsolete LAPWINK (see LAPWING), while Horny Wig appears to have a parallel in LYMPTWIGG.

Horny Wink has a secondary variant **Horry Wink**, showing local English *horry* filthy, the Lapwing traditionally having a bad reputation among country folk. Cf. HORN PIE.

Horra Goose. An Orkney and Shetland term for the Barnacle and the Brent, doubtless in the first place simply Horra, a Norn name, echoic in origin; cf. Old Norse *hurr* a dull or hollow sound, paralleled in local English; see HURROCK.

Horry Wink, see HORNY WINK.

Horse Gowk. An Orkney and Shetland name for the (Common) Snipe, a relic of the local Norn, partly Anglicized (*horse*), the ancestral form being Old Norse *hrossagaukr* literally horse cuckoo. It derives from the characteristic drumming, that resonant bleating or neighing sound heard from the flying male, especially during courtship. Cf. HEATHER BLEAT.

The idea of neighing occurs in other bird names, as HORSE MARTIN, HORSE THRUSH, also CAPERCAILLIE.

Horse Martin. An obsolete name for the Swift known from Charleton 1668 'Horse-Marten, or Swift'. The epithet *horse* derives from the squealing cry of the bird comparable to the shrill neighing of a horse; cf. synonymous SQUEALER. The notion of neighing is invoked in other bird names; see HORSE GOWK.

Horse Masher. A corruption of **Horse Match**, properly **Horse Smatch**, names for the Wheatear in use in Cornwall, otherwise SMATCH. The bird may be seen feeding on insects attracted by horse turds, hence (so we suppose) the epithet *horse*.

Horse Thrush. A Midlands expression for the Mistle Thrush. As with SHRILL COCK, SCREECH, names for this species very often allude to the shrill, screeching cry, which in the present case has been compared to a horse's neighing. A parallel is seen in one of the bird's Welsh names *caseg y ddrycin* literally mare of the storm. For examples of similar motivation, see HORSE GOWK.

House Martin. First noticed in White 1767 'house-martin, or martlet', earlier simply MARTIN, and apparently created for terminological balance after the analogy of SAND MARTIN; it was presumably inspired by HOUSE SWALLOW. Although BOU 1883 and Hartert 1912 have *Martin* as the vernacular name, House Martin was common in ornithological writing before then and has since become the accepted specific term.

House Sparrow. First in Merrett 1667 '*Passer domesticus* House-sparrow'. For Pennant 1768 'Sparrow' sufficed, but his successors generally used the present name, which with Yarrell 1843 became the undisputed standard. For local terms, see EASING SPARROW, THACK SPARROW.

House Swallow. A name for the Swallow, first in Mouffet 1655, also in Merrett 1667 and used by Ray 1678 'The common or House-Swallow'. Pennant in his writings fluctuated between this expression and CHIMNEY SWALLOW, but neither was widely used afterwards. Occurring only in the work of naturalists and not found in local use, the present term is apparently a translation of Gesner 1555 *Hirundo domestica* which, however, is based on German *Hausschwalbe*, a quite well-known folk-name.

Howl. An obsolete variant of OWL, attested first in *Owl & N. c.*1250 'hule' beside 'ule', similarly *Prompt.Parv. c.*1440 'owle, or howle'. The form shows contamination from the verb *howl*, formerly used of Owls, as Gower 1390 'The horned oule/ The which men here on nightes houle', i.e. 'which people hear howling [hooting] by night'. (*OED*). Cf. HOWLET.

Howlard. An Owl name from Ches. and Shrops., from HOWLET by suffix change; also OWLARD, OWLET.

Howlet. A local, especially northern and Scottish name for an Owl, sometimes applied to a definite species, such as the Tawny Owl. The earliest forms in two sources, both *a.*1475, are 'houlot' and 'houlat'. The present spelling dates from 1513. The word is a borrowing of French *hulotte*, fortuitously not recorded until 1530, and formed from the now obsolete verb *huler* to hoot, patently onomatopoeic.

Locally, the word may appear as **Hoolet**, more faithful to the medieval pronunciation. Where *h* is dropped, the name is heard as **Oolet**, whence the Cumb. and Yorks. variant **Yoolet**, also written **Yewlet**. See OWLET.

Howster. A name for the Knot, locality unspecified, perhaps ultimately imitative of the wheezy, whistling note sometimes heard from this bird.

Huckmuck. A Wilts. name for the Long-tailed Tit. The word occurs as early as 1492 denoting 'A strainer used in brewing. It consists of a bundle of twigs. . . (*OED*). The etymology is unknown, but there is no doubt that, as so often, the present bird name originally referred to the unique nest; cf. e.g. BAG, BUSH OVEN, FEATHER POKE.

Huplin. A Manx name for the Cormorant. It is a relic of the now extinct Manx Gaelic, in which language, however, it was a borrowing from Old Norse **hypplingr*, and thus ultimately the same word as Orkney HIBLIN. See -LING.
JMM vii 118 (1969)

Hurrock. An obsolete Scottish name for a species of Goose, known from a text of 1578 'Hurrok', as explained under ROUTHEROCK. The name consists of the onomatopoeic word *hurr* whining, wheezing, purring, hoarseness of voice (*SND*) and a suffix common in bird names, see -OCK. The species in question will have been the Barnacle or the Brent, or rather both, as indicated by the closely comparable name HORRA GOOSE.

I

Ibis, see GLOSSY IBIS.

(-ick. A diminutive suffix, comparable to the much commoner -OCK, with which it may interchange, as WILLICK beside WILLOCK; it spread to the Orkneys, see CHALDRICK under CHALDER. There are no known medieval examples.)

Ickle, see HICKWALL, also NICKLE.

Ickwall, see HICKWALL.

(-ie, also—often indifferently—written -Y. A diminutive suffix commonly occurring in pet forms, e.g. BLACKIE, CHAFFIE, THROLLY, from BLACKBIRD, CHAFFINCH, THROSTLE. It originated in Scotland and is first recorded in the surname Jamieson 1379 (*Jamie* from *James*), the earliest example in a bird name being CORBIE *c.*1420. Not surprisingly, the suffix has been especially productive in Scottish names, as LINTIE from LINTWHITE; it may acquire the function of an agent, see DORBIE. It spread not only to England, but also to Orkney and Shetland, where it modifies the Norn names, thus LYRIE beside the older LYRE, indeed these are often attested only in the diminutive form, as BAAKIE. The suffix may replace the Gaelic diminutive ending *-ag*, see STRANNIE. It may compete with the older suffix -OCK,

see WILLY (under GUILLEMOT), WRANNIE; cf. -KIE.)

Ikey, see HAY SUCK.

(-ing, -in. The suffix *-ing*, of Proto-Germanic ancestry, is employed, *inter alia*, as a means of creating animate nouns, hence also bird names, e.g. BUNTING. It has occasionally been added superfluously, see SKEELING GOOSE. In all English dialects, *-ing* developed into *-in*, the change taking place in the Middle or Early Modern periods, according to district. This spoken form has established itself in the standard language, for instance, in the case of PUFFIN, formerly PUFFING. The parallel Old Norse *-ingr* with similar functions may be traced e.g. in LORIN.

The present suffix was never prominent in the sphere of bird names, but it forms the basis of the comparable suffix -LING, -LIN, which has been much more productive, sometimes encroaching on the *-ing* type; see BUNTLIN.)

Isaac, see HAY SUCK.

Ivy Owl. A name for the Tawny Owl, recorded from places as far apart as Northumb. and Wilts. It has been known since Ray 1678 'The common brown or Ivy-Owl'. The bird may roost in ivy.

J

Jack-a-Dells. This Suff. name for the Swift will be an intentional modification of an original name *Jack Devil in order to avoid the unpleasant word DEVIL; cf. DEVILTON, DEVILING, further JACK SQUEALER, DEVIL 2, also DICKY DEVLIN.

Jack-a-Nickas, see JACK NICKER.

Jack Bird. A Worcs. name for the Fieldfare, inspired by the cry of this garrulous species. *Jack* has a comparable function in JACKDAW.

Jack Curlew. A name for the Whimbrel from the north-east coast, *Jack* implying smallness as in JACK SNIPE; cf. HALF CURLEW.

Jackdaw. Known since 1543 'Iacke dawe'. It occurs in Merrett 1667 'Jackdaw (or) Chough' and was used by Ray 1678 and Pennant 1768, since which time it has been the standard name for *Corvus monedula*. It is a compound of earlier, independent DAW with the epithet *Jack*, formally of course a nickname, but basically onomatopoeic, induced by a note often heard from this bird (cf. synonymous CHAUK), different from the clamorous cawing represented by *Daw*.

Jack has a comparable function in the Fieldfare name JACK BIRD.

Jacker. A Somerset name for a Jackdaw; see -ER.

Jack Hern. A term for the Heron, in local use from Kent to Wilts. and northwards into the Midlands, the nickname *Jack* denoting in the first place the male; the present term thus originally contrasted with MOLL HERN.

Jack-in-a-Bottle. A name for the Long-tailed Tit, the bottle being the extraordinary nest. No locality is given, but cf. widespread BOTTLE TIT.

Jack Nicker. A name for the Goldfinch chiefly in Shrops. and Ches., in the latter also **Jack-a-nickas.** In the adjacent parts of Wales, the bird is known in Welsh as *Jac Nico* and *Y Nicol* (Parry 1962), making it certain that the present names derive from Middle English *Nicol*, the then current form of *Nicholas*. The Goldfinch took this name, doubtless in pre-Reformation times, from St. Nicholas. He was popularly associated with gold through the legend of his having provided three poor girls with a bag of gold each for their dowries.

Jack Runner. A Glos. name for the Water Rail; see synonymous RUNNER.

Jack Snipe. First recorded in 1663. Pennant 1768 uses it as his main name (in preference to HALF SNIPE) and his example was followed by his successors. The epithet *Jack* refers to the bird's small size as compared with other Snipes; it has the same function in JACK CURLEW. There was a popular tradition that the Jack Snipe was the male of the Common Snipe. Such a notion will have arisen as a consequence of misinterpreting the name *Jack* as though it denoted the male bird, which it occasionally does in other cases, e.g. JACK HERN.

Jack Squealer. A Worcs. and Shrops. name for the Swift; see SQUEALER, also JACK-A-DELLS.

Jack Straw. In use for three different species: Blackcap (Shrops., Somerset), Whitethroat (Shrops., Hants.), Stonechat (Hants.). The name alludes to the straw often used in nest-building, and was surely suggested by *Jack Straw* man of straw, originally the nickname of one of the leaders of the rising of 1391. There is comparable motivation in HAY JACK.

Jacob, see GYP.

Jay. This name came to us through the Normans. It presupposes Old French *jai* (whence Modern French *geai*) beside the actually recorded form *gai*, both going back to synonymous Late Latin (5th cent.) *gāius*, which in living use replaced the Classical *grāculus*; cf. also *gāia* magpie (Classical *pīca*). The Late Latin words are evidently neologisms, basically imitative as befits such garrulous species. It follows that—whatever the appearances—this Latin name for the Jay can have nothing to do with the proper name *Gāius*, as has often been suggested. In any case, proper names were not used to denote birds until the Middle Ages, see under MARTIN.

The name Jay makes its debut *a*.1310. It seems to have established itself quickly and is used by all the early naturalists. It has quite replaced Old English *higera*, cognate with German *Häher*, also ultimately onomatopoeic.

Jay Pie, in the north and Scotland **Jay Piet**. Expansions of JAY, fairly widespread in their respective areas, the Jay being classed with the Magpie.

Jebbie, see GYP.

Jenner Hen. A Yorks. variant of the more usual JENNY WREN. The first word shows the influence of **Chitter Wren** or a similar form, see further under GILLER WREN, while the substitution of HEN for WREN is due to the once well-known rhyme 'The Robin and the Wren/Are God Almighty's Cock and Hen.'

Jenny Crudle. A Yorks. name for the Wren, *crudle* possibly meaning literally a small lump.

Jenny Wren. A widespread, familiar term for the Wren, making its debut in 1638 as 'Jinny Wren' in accordance with the alternative pronunciation. *Jenny/Jinny* are pet forms of *Jane*, added to the original name, as in JUGGY WREN, also KITTY WREN. Sometimes abbreviated to **Jenny/Jinny**. See JENNER HEN.

Jer Cock, see CHAR COCK.

Jester Bird. A name for the Starling, locality unspecified. Apparently a

further corruption of CHEPSTER; see under SHEPSTARE.

Jet Cock. A term first found in Tunstall 1771 'Jet or Jud-cock or Jack Snipe', to all appearances a folk-etymological development of JUD COCK.

Jobbin, Jobin, see NUT JOBBER.

Joe Ben. An East Anglian name for the Marsh Tit or the Great Tit, further a Glos. variant **Joe Bent** meaning the Great Tit. We suppose that the latter sense and the form Joe Ben (i.e. the biblical brothers) are primary, and suggest that this two-syllable name will ultimately be a reflection, however devious, of the arresting and quite unique two-syllable call of the Great Tit, as so often in names for this species; see further under SAW SHARPENER.

Joey, see GYP.

Jud Cock. A name for the Jack Snipe, reported from half a dozen counties, southern as well as northern. Attestation goes back to 1621 'jude-cocks' or 'Jugcocks', the latter leaving a trace in Notts. **Juggy**. The origin and literal meaning of the element *jud-*, *jug-* are unknown. Cf. JET COCK.

Juggy Wren. A Surrey name for the Wren. *Juggy* (and *Jug*)—now obsolete: Shakespeare (*Lear*) 1605/6 'Whoop, Jug! I love thee'—were popular pet forms of *Jane*. The present name is thus particularly close to JENNY WREN. Other examples of the use of this pet name in BOTTLE JUG, HEDGE JUG, JUG POT.

Jug Pot. A Notts. term for the Long-tailed Tit. The *pot* is the remarkable nest, and *Jug* a pet name, as explained under JUGGY WREN. The name is closely comparable to BOTTLE JUG.

Julliver Wren, see GILLIVER WREN.

Juniper Titmouse. An obsolete term for the Crested Tit, apparently a nonce word due to Charleton 1688 '*Parus cristatus*, the Crested or Juniper Titmouse'. This species may take juniper berries.

K

Kae, also spelt **Kay.** A northern, today usually Scottish, name for the Jackdaw. It is quoted by Turner 1544 'monedula . . . a caddo, a chogh, a ka', the record going back to c.1325 'co'. It is likely that the ultimate source is Old Norse *ká presupposed by Norwegian *kå*. At any rate, the name is clearly echoic, inspired by the noisy cawing. Cf. DAW.

Katogle. A Scottish name for an Owl, apparently of Norn origin going back to Old Norse *kattugla, attested in the derivative languages, e.g. Norwegian, literally cat owl, the name originally denoting the Tawny Owl, from the cat-like appearance of the head. Sometimes Anglicized as **Cat Owl.** Cf. CATA-FACE, YOGLE.

Keelie. A Scottish and Northumbrian name for the Kestrel, from the clear, penetrating call heard during courtship, interpreted as *keel*, with the locally prolific suffix -IE.

Kentish Crow. An East Anglian name for the visitant Hooded Crow. Here *Kentish* virtually means foreign. Cf. NORWAY CROW.

Kentish Plover. This rare species, which the layman would not distinguish from the Ringed Plover, was first named by Latham 1802, specimens having been sent him from Sandwich in Kent.

Kepshite. Given me by Mr. J. Simpson as a Northumbrian term for the Arctic Skua, literally catch-shit; for motivation and comparable names, see under DIR-TEN ALAN.

Kestrel. The name occurs in two main variants, both in Turner 1544 'kestrel, kastrel'. The latter, often more naturally spelled **Castrel,** survived locally well into the last century. Ray 1678 used the former variant and Pennant 1768 followed suit, though spelling 'Kestril'. His successors sometimes imitate this, but the present spelling became more general with Yarrell 1843, though he himself could also use 'Kestril'.

The name is French, in which language it has been recorded in a medley of forms, *crécerelle* emerging as the standard. The variant closest to English is preserved in Poitevin dialect *casserelle*, which would give in earlier English *casserel, easily contracting to *casrel, whence with excrescent *t* the recorded CASTREL. The form Kestrel may be regarded either as a reflex of an unrecorded French variant *quesserelle*, or as a regional development of **Castrel** showing the same vowel change as seen e.g. in local *peth* for *path*. See -EREL.

The name is doubtless of onomatopoeic origin, alluding to the courtship call (cf. KEELIE) and apparently cognate with *crécelle* (a) rattle. The two are often confused, so that *crécelle* can also denote the bird. See CRESS HAWK, CRISTEL HAWK.

For an earlier, native English name, see STANNIEL, cf. also WINDHOVER.

Ket Crow. A Yorks. (West Riding) name for the Carrion Crow, *ket* meaning carrion. The name is thus the local equivalent of CARRION CROW. It appears in Lincs. as **Cad Crow.** Cf. GORE CROW.

Kiddaw. A name for the (Common) Guillemot, as taken down at St. Ives, in Ray 1674. It is Cornish *kidda*, imitative of the shrill cry of the juvenile bird. Cf. SKIDDAW.

(-kie. A Scottish suffix amalgamating two diminutive elements, -k- (see -OCK) and -IE; it occurs in DOVEKIE, WILKIE.)

Killigrew. A name for the Chough occurring in Charleton 1668 'Killegrew', presumably for *Killigrew Crow** after the Cornish village of that name; cf. in principle MARKET JEW CROW. The name is said to have been in popular use in Cornwall.

Killy Leepie. A Scottish name for the Sandpiper, imitative of the shrill call, with the locally productive suffix -IE; cf. KITTY NEEDIE.

King Duck, King Eider. Pennant 1785 described this species under the name King Duck, a term still used by Yarrell 1843 and other editions. Meanwhile Fleming 1828 had introduced King Eider, which in the latter part of the century eclipsed Pennant's name: BOU 1883, Hartert 1912 'King Eider'.

The King Eider is an Arctic bird, most likely to be seen in winter when the birds move southwards. A single specimen may then associate with a flock of Common Eiders. Should the bird in question be a male, his striking appearance marks him out from all the rest: he is the king, as expressly stated in a report from the Faroes quoting the native name *æðukongur*, similarly Icelandic *æðarkóngur* (Lockwood 1961, pp. 11f.). This Scandinavian tradition is, of course, the inspiration of Fleming's name.

Kingfisher, formerly **King's Fisher.** The oldest known name for this bird is found in Old English *īsen, īsern*, considered to mean iron-coloured i.e. blue; it is of West Germanic age, and survives corruptly in German *Eisvogel*. But as early as *c*.1000 the name *fiscere* was also in use, and although not explicitly recorded during the Middle English period, conceivably lives on in local Yorks. **Fisher.** The now current Kingfisher, apparently to be interpreted as excellent fisherman, does not make its debut until 1658, and is moreover preceded by King's Fisher, first attested as a personal nickname in 1318 'Henry le kingesfisscher'. Pennant 1768 lent his authority to Kingfisher, but King's Fisher was still used by Selby 1833.

The name is reminiscent of local French *roi pêcheur*. It looks as though the royal epithet came to us from that source, where we surmise the name to be a reflex of Old French *Roi Pescheor* Fisher King, familiar from the Grail Romances and going back to the second half of the 12th cent.

King Harry. A name for the Goldfinch, found here and there in various parts of the country. This looks like Henry VIII, that monarch's ostentatious attire being linked with the ornate plumage of the bird. Cf. PROUD TAILOR, SHERIFF'S MAN.

Kinglet, see GOLDCREST.

Kipp. A Kent name for a Tern, imitative of the harsh scream, like most names for these birds; cf. SKIFF and see further under TERN.

Kirr Mew. This name for a Tern (locality not specified) preserves in MEW the old word for a Gull. The first element is an imitation of the penetrating cry, and has a close parallel in SKIRR; see also TERN.

Kite. The name goes back through Middle English *kyte* to Old English *cȳta*, the latter glossing *Buteo* buzzard. The name is imitative of the long-drawn, whistling call common to both Kite and Buzzard, and may be found locally in use for the latter species too. Having no cognate in other Germanic languages, the term most likely arose in this country, apparently originally denoting the Buzzard only, the Kite at any rate already having a well-known Germanic name in GLEDE. In the event, the name came to be attached to *Milvus* too, and was so used by the early naturalists, as Merrett 1667 'fork-tailed Kite', Ray 1678 'Kite or Glead', Pennant 1768 'Kite', whence modern standard usage. For another traditional name, cf. PUTTOCK. See BALD KITE.

The name of the toy, known since 1664, is a special application of the bird name, suggested by the soaring flight.

Kittiwake. First in Ray 1661 (misspelt) 'cattiwake', next Sibbald 1684 'Kittiwake'. Pennant 1768 took up the name, thus giving it in practice standard status. The name is Scottish. Ray and other early authorities associate the word with the birds on Bass Rock, so that quite possibly the name arose in that area. It is echoic, after a commonly heard three-note cry. Kittiwake names which have remained local include ANNET, HACKET, RITTO, WEEO.

Kitty Longtail. A Somerset name for the Long-tailed Tit.

Kitty Needie. A Scottish Sandpiper name, basically onomatopoeic and closely comparable to synonymous KILLY LEEPIE.

Kitty Wren. A popular name for the Wren, fairly widespread, especially in the north, *Kitty*, a pet form of *Katharine*, being prefixed as in JENNY WREN.

Perhaps the name was suggested by CHITTY WREN or TITTY WREN.

Knat. A Norf. variant of KNOT, known from *a*.1616. A further variant **Gnat**, from the same county, was recorded by Browne *a*.1682 'Gnats or Knots'. All these forms are imitative of the grunting call, both *g* and *k* being originally pronounced, evidently still the case in Norfolk in the late 17th cent.

Knot. First noted in 1452 'Knottys', the present spelling in Drayton 1622, similarly in Ray 1678 and Pennant 1768. It is imitative of the grunting note, and the *k* was widely pronounced until the 18th cent. Cf. KNAT.

L

Lairblade, Lairbladin. Orkney terms for the Cormorant, relics of the former Norn language, presupposing respectively Old Norse *lærblettr* thigh spot and *lærblettingr* bird with the thigh spot, in allusion to the white patch on the Cormorant's flanks, the second element being modified under the influence of English *blade*. The same motivation underlies the synonyms HIBLIN and LORIN. See -ING, and also SHAG.

Land Rail. First in Ray 1678 'Land-Rail', a term for the Corncrake coined after Latin *Rallus terrestris* (recorded since the 13th cent.). Pennant 1768 adopted the name, but subsequently abandoned it in favour of CRAKE GALLINULE, a term taken up by most of his immediate successors and still the main name in Montagu 1813, but then dropped. Gould 1837 restored the present term. Yarrell 1843 took it as the principle name in preference to CORNCRAKE, introduced into the mainstream of ornithological literature by Bewick 1797. But it later declined in spite of Newton 1896 'The Land-Rail, also commonly known as the Corn Crake, sometimes the Daker Hen'. In this century Land Rail, though not uncommon, has largely given way to its chief competitor and is now the less usual term. See WATER RAIL.

Langy. A Hebridean name for the Common Guillemot; it is Gaelic *langaidh*, a loan from Old Norse *langvíi*; see under LONGIE and cf. LAVY.

Lappinch. A Ches. word for the Lapwing, a modification of obsolete LAPWINCH (see LAPWING), doubtless to remove the intolerable element -*winch*.

Lapwing, also obsolete **Lapwinch, Lapwink.** By singular good fortune, the name is attested in (the Mercian dialect of) Old English in three instances from the 8th and 9th centuries: 'lǣpiwince, lǣpewince, lēpewince'. These forms reflect the expected phonetic evolution of the dialect during the period concerned and are therefore entirely reliable. The first element 'lǣpi-' was evidently once a bird name in its own right, for it is etymologically identical with West Frisian *leep*, North Frisian *liap* lapwing, further Heligoland Frisian *leap* hoopoe, presupposing Proto-Germanic *laipiz*. The Heligoland sense is secondary; what these species have in common is the remarkable crest, and crest is the basic meaning of this word. As a bird name, it is thus an illustration of the part-for-whole principle; see under SHAG.

The second element -*wince* is formally winch, a word having the primary sense of something moving up and down, which in the bird name must have referred to the movable crest. It now appears that, in Old English, the literal meaning of *lǣpi* was forgotten and came to be reinforced by another word for crest, the part for whole principle repeating itself. Lapwing names are often motivated by the crest; see WIPE and HORNY WICK.

But this name, so reliably attested in early Old English, was soon subject to numerous corruptions, as seen in the next two attestations, both from West Saxon sources of the 11th cent.: 'hlēapwince' and 'lapawinca', the former showing contamination from *hlēapan* to run, leap, the latter with *lapa* presumably standing for *lappa* lappet and *winca* winker. Such folk-etymological distortions continue into the Middle English period: 1340 'lhapwynche', 1390 'lappewinke', 1430 'lapwingis' with the first appearance of -WING, 1481 'lapwynches' surviving locally in a modified form, see LAPPINCH. By the 16th cent., however, LAPWING had become usual, but there is the exceptional 1569 'lapwink', with which compare HORNY

WINK. See FLOP WING, LIPWINGLE, LYMP-
TWIGG.

Especially in the north of England and
in Scotland, the present name type is
replaced by onomatopoeic formations
(see PEWIT and TEUCHIT) but Pennant
1768 preferred Lapwing, which thus
became the regular term in ornithology.
In popular use, the bird is often called
GREEN PLOVER.

Lark. A Germanic name: Middle English
larke, Old English *lāwerce*, older *lāwrice*,
ultimately identical with Dutch *leeu-
werik*, German *Lerche*; to these, Old
Norse *lævirki* is very close. We postulate
Proto-Germanic **laiwizikō*, where the
ending *-ikō* contains a diminutive *-k-*
suffix and *laiwiz-* is the stem of an
onomatopoeic word for song. The name
then literally means little song, in
modern idiom little songster, and to this
extent is comparable to SWAN. See
LAVEROCK.

The simplex Lark denotes first and
foremost the Skylark, and this use is
found in ornithological works before
Ray 1678; see SKYLARK. More loosely, in
local usage, Lark includes both the
Skylark and the Woodlark, traditionally
also the Pipits. It may appear in other
names, notably SEA LARK and SAND
LARK.

Fróð. xxx 103–5 (1982)

Laughing Bird. A Shrops. expression
for the Green Woodpecker, motivated
by the call. Cf. YOCKLE.

Laughing Goose. A fairly widespread
name for the White-fronted Goose,
from its loud, laughing cackle.
Although traditional and used (the first
record) by Edwards 1753, it was not
taken up by later naturalists.

Laverock. A widespread, semi-literary
equivalent of LARK, dominant locally in
the north and in Scotland. It goes back
to Middle English, when it was also a
literary word. It has been reshaped
from Old English *lāferce*, a secondary
variant of *lāwerce* (see LARK), by the use
of the diminutive suffix -OCK.

Lavy. A Hebridean name for the Guillemot, first in Martin 1698 who reports
the name from St. Kilda. It is Gaelic
làmhaidh, in which language it is a
borrowing from Old Norse **lamvī* (cf.
Faroese *lomvigi*) from earlier **langvī*,
see under LONGIE and cf. LANGY.

(-le. An agent suffix going back to Middle
and Old English -*el*, cognate with Old
High German -*il*, further Old Norse -*ill*,
all presupposing Proto-Germanic **-ilaz*.
It is attested e.g. in Old English *pyttel*
kite (or similar species) which survives
corruptly in DUN PIDDLE, DUN PICKLE. A
suffix based on the consonant *l* may, in
very ancient names, have a diminutive
or hypocoristic function, as in THROS-
TLE, ultimately also in QUAIL, and
apparently the same in OWL.)

Leach's Petrel. A species identified by
W. E. Leach in 1817 and found the
following year by Bullock to be breeding
on St. Kilda, hence Temminck 1820
'*Procellaria leachii* Pétrel de Leach', first
in English in Jenyns 1835. This name
competed with FORK-TAILED PETREL (introduced 1828) and both terms remained in use for the rest of the
century, witness Newton 1896 'Leach's
or the Fork-tailed Petrel'. Meanwhile
the conflict has been resolved thanks to
a refinement in terminology. Thus Hartert 1912 combined the two expressions
to give **Leach's Fork-tailed Petrel**
(*Oceanodroma leucorrhoa*) enabling him
to distinguish the very closely related,
and rarest of vagrants, the 'Madeiran
Fork-tailed Petrel' (*O. castro*). These
cumbersome names were sometimes
followed, but latterly the perfectly
adequate Leach's Petrel (cf. 'Madeiran
Petrel') has again become usual, thus
Checklist 1952, as in BOU 1883.

Lennet. A form of LINNET, occurring
sporadically in various parts of the
country, first noticed in 1604. Also with
change of suffix Glos. and Oxon. **Len-
nard**, north country **Lennart**.

(Lesser, occasionally **Less.** An epithet
introduced by the early ornithologists,

translating Latin *minor*. However, they may also employ LITTLE in the same sense, often using both indiscriminately in the same work.)

Lesser Ash-coloured Heron. A term for the Night Heron coined by Ray 1678 'The lesser ash-coloured Heron, called by the Germans, the Night-raven'.

Lesser Black-backed Gull. Coined by Montagu 1802 as 'less Black-backed Gull', but taken up by subsequent authors in its present form. See GREATER BLACK-BACKED GULL. For a folk name, see SADDLE BACK.

Lesser Grey Shrike. A term created by Pennant 1785.

Lesser Red-headed Linnet, see RED-HEADED LINNET.

Lesser Redpoll. Coined by Montagu 1802 'Lesser Redpole' and accepted by virtually all succeeding writers. The spelling has vacillated; see REDPOLL. Cf. GREATER REDPOLL.

Lesser Spotted Woodpecker. The name is due to Ray 1678. Pennant 1768 modified this to 'Lest (i.e. Least) Spotted Woodpecker', but his successors preferred the present form. See BARRED WOODPECKER.

Lesser Whitethroat. A term coined by Latham 1787 to name the species first separated from the Common Whitethroat by Gilbert White in 1768.

Lich Fowl. A Ches. and Shrops. name for the Nightjar, literally meaning corpse bird, the uncanny, nocturnal bird being associated with death. The name must go back at least to the late Middle Ages when *lich* was still a living word and FOWL the ordinary term for a bird.

(-ling, -lin. The suffix *-ling*, found in both West and North Germanic languages, consists of an *l* element added to an original suffix *-ing* (see -ING). Among our English examples, STARLING is the oldest, going back to late Old English times; none of the others are likely to

have developed before the Middle English period when the suffix became particularly productive. In its earliest function it denoted the young bird. Later, it was employed to indicate association with a habitat, as REEDLING, or referred to the colour of the feathers, as BLAKELING (*blake* yellow). It could be added to verbal forms; see OTTERLING. Finally, it may occur as a redundant appendage, as in SANDERLING. The parallel Old Norse had comparable functions; see TITLING.

In later spoken English, *-ling* is commonly pronounced *-lin*, and local names are often recorded in this form; it has passed into the standard language in the case of DUNLIN.)

Linnard. A variant of LINNET arising by suffix substitution; it has been recorded in Hants. and Wilts., further Cumbrian **Linnart**.

Linnet. A borrowing of Old French *linette*, first in English *c*.1530; the word has been known in French since the 13th cent. It consists of *lin* flax going back to Latin *linum*, and the diminutive suffix *-ette* here having the force of associated with, thus comparable to our English suffix -LING in a name like REEDLING—it goes without saying that the Linnet is partial to linseed. In popular use, the present name commonly includes the Twite and the Redpolls, these various species not being readily distinguished by the layman. It was Pennant 1768 who effectively established the term in its present specific sense, Ray 1678 having referred to it as 'The common Linnet' in contradistinction to 'Mountain Linnet' (i.e. Twite), etc.

The form Linnet is often modified in local use, cf. LINNARD, LENNET.

For older Linnet names, see LINTWHITE, THISTLE TWEAKER.

Lintie. A common name in Scotland for the Linnet, from *lint-* in LINTWHITE with the hypocoristic suffix -IE. The record goes back to 1728.

Lintwhite. A term for the Linnet, best known from Scotland, where it apparently arose; the first attestation dates from *c*.1600 'Lint-white'. It is preceded by two Middle English occurrences (approximately first half of the 15th cent.) 'lyngwhittes' and 'lynkwhyttez', i.e. normalized singular *lingwhīt, linkwhīt*. All three variants are plausibly folk-etymological reformations of Old English *līnetwige (līne-* flax, *-twige* plucker; cf. THISTLE TWEAKER). In the present name, *lint-* is a Scottish word for flax, whence also the local adjective *lint-white* white as flax. Cf. LINTIE.

Lipwingle. Bedfords. variant of LAPWING. The first syllable was perhaps intended to imitate the bird's call, the second is presumably to be explained from the Derbys. expression *wanglewingle* shaky, with reference to the movable crest.

(Little. This epithet occurs in a few folk names, e.g. LITTLE SNIPE, LITTLE WHAUP, but the other instances—all standard names—are artificial coinages, often ultimately reflecting Latin *minor*, as explained under LESSER.)

Little Auk. Coined by Pennant 1768, an Anglicization of *Alca minor* abstracted from Linnaeus 1758 *Alca alle* and Brisson 1760 *Uria minor*; see further under AUK. Cf. ROTCHE.

Little Bustard. A term introduced by Edwards 1758 to distinguish this species, only rarely encountered in Britain, from the much larger Great Bustard, at the time a resident bird.

Little Egret, see EGRET.

Little Grebe. Coined by Pennant 1768 for systematic reasons and used by succeeding authorities until Yarrell 1843 'Little Grebe or Dabchick, as it is more generally called'. After that, the artificial term and the folk name competed in the literature, the former being preferred in more technical contexts, as BOU 1883, Hartert 1912, *Checklist* 1952.

Latterly, however, it has become more usual to separate this species generically from the other Grebes, so that the present term has lost something of its original validity and has accordingly lost ground to DABCHICK.

Little Owl. Due to Ray 1678 and used by all subsequent writers. Though common enough today, the Little Owl is an introduced species, the first attempted introduction taking place in 1842. It was previously barely known, being then only a rare visitor without a recognized English name, hence Merrett 1667 could only translate and describe 'Noctua Night, or little grey Owl'.

Little Ringed Plover. Although a British nesting species since the 1930s, this bird is still rare and was more so in the last century. Not surprisingly, it remained unnoticed until the age of scientific ornithology and without a popular name. The present term was coined by Jenyns 1835 and taken up by Yarrell 1843, since which time it has been part of the standard nomenclature.

Little Sandpiper. A term for the Little Stint coined by Pennant 1768 and commonly used by the naturalists until Yarrell 1856 decided in favour of LITTLE STINT.

Little Snipe, see STINT.

Little Stint. First met with in Bewick 1797 and adopted by Jenyns 1835. Yarrell 1843 had 'Little Sandpiper, or Little Stint', but in 1856 gave precedence to the present term, since which time it may be regarded as the standard.

Little Tern. First noticed in Gould 1834, conceivably suggested by Linnaeus 1758 *Sterna minuta*. Pennant 1768 and succeeding authors, including Yarrell 1843, generally call this bird the 'Lesser Tern', but the present name came to the fore in the latter part of the 19th cent., appearing in BOU 1883.

Little Whaup. Literally little Curlew, a local Scottish name for the Whimbrel, first recorded in 1795. Cf. HALF CURLEW.

Long-eared Owl. Coined by Pennant 1768 and generally accepted; the name contrasts, of course, with SHORT-EARED OWL.

Longie. A Shetland name for the Guillemot coming down through Norn from Old Norse *langvíi (so attested in Icelandic langvíi), literally long víi, the latter element onomatopoeic, based on the call of the juvenile (cf. WEEO), in which connection see further WILLOCK (under GUILLEMOT), and compare LANGY, LAVY. Fróð. xxii 107–10 (1974)

Long Pod. A term for the Long-tailed Tit from Surrey and Somerset. Pod has here its former meaning of bag, the name denoting in the first place the long, bag-like nest. The word pod was originally *pud, the present form being due to contamination with synonymous cod (the sense still in peascod); cf. POKE PUDDING.

Long-tailed Duck. A term coined by Edwards 1750 and used by practically all succeeding authors. Folk names include CALLOO, COAL-AND-CANDLE-LIGHT, COLDIE, DARCALL, also SEA PHEASANT.

Long-tailed Tit. Generalized by Yarrell 1843, for older **Long-tailed Titmouse** used e.g. by Pennant 1768; it makes its debut in ornithological literature in Merrett 1667 'Parus caudatus least, or long-tail'd titmouse'. The name, actually first encountered in Cotgrave 1611 'mesange à longue queuë long-tayled Titmouse', is a book name, taken up in preference to any of the numerous popular ones. Some of these likewise refer to the tail, as KITTY LONGTAIL, but the majority allude to the extraordinary nest, e.g. BOTTLE TIT, in fact most of them properly denote the nest itself, hence BAG, BUM BARREL, BUSH OVEN, FEATHER POKE, HUCKMUCK, JUG POT, etc. See also under WILLOW WARBLER.

The present term was formerly systematically appropriate, the species being for long ascribed to the genus Parus.

Loo. A Shetland name for the Ringed Plover, mostly in the form **Sandy Loo**, from the local Norn, a more archaic form being seen in Orkney SANDLO. In the variant (Dunrossness) **Sandy Loog**, the present word has developed a final g, a feature paralleled in the closely related Faroese lógv beside original ló, dating from between 1400 and 1600; cf. SHOOG. The ultimate source is Old Norse ló, of onomatopoeic origin.

Loon. Literally fool, a local name for a Diver or, less usually, for a Grebe. In the former sense it occurs first in 1634 'Loone', in the latter in Ray 1678 'The greater Loon or Arsfoot'. It is assumed that this name, chiefly northern and East Anglian, is a corruption of Old Norse lómr red-throated diver. The primary sense is moan i.e. moaning bird, as shown by the two meanings attested in Icelandic lómur red-throated diver; moan. The allusion is to the loud, wailing call, as commonly in the case of Diver names; cf. EMBER GOOSE. Loon is also the standard North American name for the Great Northern Diver. On the ancient use of an abstract noun to form a bird name, cf. SWAN.

Lorin, also **White Lorin.** Norn terms from Shetland for the Cormorant. They presuppose Old Norse *lāringr, abbreviated from *hvītlāringr (likewise implied by Norwegian kvitlåring) literally bird with the white thigh (lār), an allusion to the conspicuous white patch which develops on the bird's flanks during the breeding season. The same motivation is seen in HIBLIN, LAIRBLADE. See -ING.

Loriot. An obsolete, bookish word, apparently not always exactly understood, but properly denoting the Golden Oriole. It occurs first in 1601 'the Witwall or Loriot' and was still occasionally used in literary work in the late

19th cent. The name has been lifted direct from French *loriot*, known since the 14th cent.; it stands for *l'oriot* the *oriot*, the definite article becoming part of the later form. Old French *oriot* itself arose by suffix change from *oriol*; see ORIOLE.

Lymptwigg. An Exmoor variant of LAPWING, known from a glossary printed in 1746. The first element *lymp-* is simply a spelling of limp. The second appears to have originally been *-wigg*, paralleled in the synonym HORNY WIG from neighbouring Cornwall (see under HORNY WICK), the medial *t* developing later for phonetic reasons, much as *d* in YOULDRING, earlier YOULRING.

Lyre, Lyrie. Shetland and Orkney words for the Manx Shearwater, from the former Norn language. They denoted in the first place the plump and fleshy nestling—the adult by contrast is tough and lean, see SCRABER—as shown by the Shetland saying 'fat as a lyrie' and confirmed by the etymology. Lyrie is the diminutive (formed with Scottish *-IE*) of the primary form Lyre descended from Old Norse *līri* nestling shearwater, the ultimate meaning of which is simply fat (noun).

Lockwood 1961, pp. 32–7.

M

Maa, see MAW.

Maddrick Gull. A name from Cornwall for the Black-headed Gull, the original form being, presumably, simply **Maddrick**. Affinities unknown, but almost certainly a relic of the Cornish language.

Madge. A name for an Owl, especially the Barn Owl, first occurring in 1591. It has been reported from Warwicks. to Norf. and Worcs. This pet form of the Christian name *Margaret* originally qualified OWL, etc., although—fortuitously—the actual attestations are of more recent date: 1598 **Madge Howlet**, 1637 **Madge Owlet**, 1823 **Madge Owl**. Cf. MOGGY, PADGE.

Mag. In local use for MAGPIE, of which it is an abbreviation; cf. MAGGET, also MIGGY.

Magget, older spelling **Magot**. A name for the Magpie, from Lincs., Glos. and Wilts., first noticed in 1791. It is an abbreviation of MAGOT PIE; cf. MAG.

Magot Pie. A 17th cent. name for the Magpie, first found in Shakespeare (*Macbeth*) *c.* 1605–06 'magot pies'. The expression derives from an original *Magot* (for **Margot*) *the pie* (1573 'magget the py') modelled on French *Margot la pie*, of which the English record is (vicariously) the first attestation. A further variant is seen in 1598 'maggot a pie', whence Cotgrave 1611 'magatapie' and present-day **Maggoty Pie** heard in Worcs., Glos., Wilts.

Magpie. This name can hardly be much older than its first attestation in 1605 'as merie as a magge pie' (*OED*); nonetheless it spread quickly, offering an easy alternative to the ambiguous PIE. Apparently at first a Midlands or southern word, it contrasted with the northern PIANNET, cf. Ray 1678 'Magpie, or Piannet', but with Pennant 1768, it became the undisputed standard name.

It derives from MAGOT PIE; see also PIE MAG, NAN PIE.

Mall, see MAW.

Mallard. The record goes back to 1314 'mallard', but until recently the word was more often used to designate the male of this species only. This sense first occurs *c.*1330 'maulard'; it is evidently the original one: the name came to us with the Normans, cf. Old French 12th cent. *malard* mallard drake. The word contains the familiar suffix -ARD, but nothing further is known about its origin or affinities.

In general, the early naturalists preferred to use the present term to mean the drake only, their name for the species being WILD DUCK. Pennant 1768, however, used the present term specifically, but in this he found few followers; from Yarrell 1843 to Newton 1896 Wild Duck remained the specific term. But Hartert 1912 went back to Mallard which has since become the recognized standard, the male being now called the Mallard Drake.

Mallemuck. A name for the Fulmar, first found in F. Martens, *Voyage to Spitzbergen (1671)* 1694 'Mallemucke'. It is Dutch *mallemok*, i.e. *malle* foolish, *mok* gull, alluding to the folly of the birds in allowing themselves to be easily taken; cf. our FOOLISH GUILLEMOT. The name, which is mainly in Scottish use, appears to have spread from Orkney and Shetland, where it would be picked up from Dutch sailors calling there on their way to and from northern waters. The species in question became a familiar British bird only within the last hundred years and never possessed a native English name; see FULMAR. In these circumstances, it is not surprising that such an exotic term as the present one should have been taken into the language.

For a comparable borrowing, see
ROTCHE.

Malpe. A name for the Bullfinch, attested
in a document of 1673 from Lancs. (*EDD*
under Mawp). Apparently a cross between MAWP and ALP, two other names
for this bird.

Manx Shearwater. The Calf of Man was
formerly this Shearwater's chief stronghold in Britain, the breeding colony
being probably the largest ever known;
it was destroyed by rats about 1800. The
bird was first described in literature by
Ray 1678 under the name 'Puffin of the
Isle of Man', in the index called 'Manks
Puffin'. This Pennant 1768 altered to
'Manks Petrel', but in 1776 he substituted 'Shearwater', on the basis of
which Selby 1883 introduced the present term as the specific name of
Puffinus puffinus, succeeding authors
following suit.

Marble Thrush. A Northants. name for
the Mistle Thrush, suggested by the
dark blotches on the breast.

Marigold Finch. A name for the Goldcrest, locality unspecified.

Market Jew Crow. A name for the
Chough, Market Jew being an alternative name for Marazion (near Penzance)
where the birds apparently congre-
gated. Both forms of the village name
are corruptions of the same original
Cornish *Marghas Dēth-Yow* literally Market (of) Thursday.

Marrot. A Scottish name for the Guillemot, imitative of the guttural call of the
adult bird; cf. MURRE. The name, known
since 1710, is often transferred to the
Razorbill, a species which commonly
consorts with the more numerous Guillemot.

Marsh Harrier. Early naturalists used
the folk-name MOOR BUZZARD to denote
this species, but Edwards 1760 introduced **Marsh Hawk**, apparently his
own creation. Selby 1825 took this
latter, more suggestive name, but

changed Hawk to Harrier for systematic
reasons. Yarrell 1843 adopted Selby's
term, which has since been regarded as
the standard.

Marsh Tit. Generalized by Yarrell 1843,
for the older **Marsh Titmouse**, first met
with in Ray 1674, whence 1678 'Marsh
Titmouse, or Black-cap', with which
compare **Fen Titmouse** occurring in
Charleton 1668 'Black Cap, or Fen
Titmouse'. Such names are unexpected,
seeing that the species is a woodland
bird. They are, however, not original
English, but the respective naturalists'
translations of Gesner 1555 *Parus palustris*, the folk-name being BLACKCAP.
Pennant 1768 copied Ray's names, but
in 1776 he put 'Marsh Titmouse' only,
effectively making it the standard term
for the present species, Blackcap being
reserved for (what is now termed) *Sylvia
atricapilla*; see BLACKCAP.

Marsh Warbler. A name which may be
said to date from 1871 when the bird
was first identified as a British species,
distinct from the Reed Warbler. It is a
translation of the older scientific term
Sylvia palustris; see WARBLER.

Martin. The first attestation dates from
Holland *c*.1450 (in the typically Scottish
form) 'Martoune', the second 1589
'Martin', then 1591 'marten', next Minsheu 1627 'martin, marten, martinet,
martelet', the spelling 'marten' again in
Charleton 1668 (see HORSE MARTIN),
further Ray 1678 'Martin, or Martinet,
or Martlet', Pennant 1768 'Martin'.
From Ray on, the literary meaning is
unequivocally House Martin, but the
sense Swift survives locally.

French *martin* is also used as a bird
name, denoting both the House Martin
and the Swift, as well as other birds,
particularly a species described by Buffon 1775 as being in size between the
Blackbird and the Starling, hence ultimately the modern standard term *martin roselin* rose-coloured starling. The
name first appears in the records in
martin-pêcheur kingfisher, since 1680,

synonymous with *martinet-pêcheur*, recorded in the previous century, see MARTINET.

The present term is, of course, a special application of the Christian name *Martin* which came to us with the Normans, becoming (from the 12th cent. on) a great favourite here, as in France. The practice of bestowing proper names on creatures of the wild is traceable to 1112 when, in Flanders, *Ysengrim* is recorded as the name of the wolf who figures in the Beast Fables. This was an innovation, the earlier tradition having no such names. The Fables are known to have been very popular in the 12th cent., when the present bird name may well have arisen; at any rate the form Martinet is recorded from the end of that century. See also MARTLET, HOUSE MARTIN, SAND MARTIN.

Before the adoption of the French name, the Martins would have been known as Swallows, as in Dutch *huiszwaluw*, *oeverzwaluw*, German *Hausschwalbe*, *Uferschwalbe* literally house swallow, bank swallow, with which compare certain local English terms such as EASING or WINDOW SWALLOW for the House Martin, BANK or SAND SWALLOW for the Sand Martin.

Martinet, also contracted **Martnet.** An obsolete name for either the House Martin or the Swift, first coming to notice in *Prompt. Parv. c.*1440 'martinet, martenet', next 1513 'martynet', the same in Palsgrave 1530, then Minsheu 1627 'martin, marten, martinet, martelet'. With Ray 1678 'Martin, or Martinet, or Martlet', the name specifically denotes the House Martin. Pennant 1768 used 'Martin' (for the House Martin) and 'Swift', thus ignoring the present term which eventually fell into disuse, the *OED*'s last citation being Montagu 1833 'Martinet. A name for the Window Swallow'.

The name in question is from French, but in that language may also refer to other birds, notably the Kingfisher, in which sense it makes its debut, in Latin dress and in the feminine gender, as *martineta* in Giraldus 1185. Almost contemporaneously, a Latin document from Cambrai *c.*1200 applies the name to the Blackbird. The first attestation in a French-language text, a 14th cent. source (Auvergne dialect) uses the same name to denote the House Martin. French *martinet* (masculine), known since Palsgrave 1530, most commonly designates the House Martin or the Swift, the latter sense prevailing in the standard language, the term for House Martin being *hirondelle de fenêtre* comparable to our WINDOW SWALLOW. The word is a diminutive of *martin*, with which it may be interchangeable; cf. now unused *martinet-pêcheur* kingfisher from 1553 with its modern synonym *martin-pêcheur*. See MARTLET.

Martinet is the oldest known example, by far, of the use of a Christian name to denote a bird; see further under MARTIN. Since such names are applied to familiar or attractive species, it is not surprising that here the diminutive, i.e. the pet form, has from the beginning been very much in evidence. Indeed, if the analogy of French *sansonnet* starling (properly **samsonnet* from *Samson*) or of such English names as JENNY WREN (with pet form of *Jane*) be any guide, the diminutive may well be the primary form. Cf. ROBINET.

Martin Oil. A Galway name for the Stormy Petrel, a Gaelicism presupposing *Máirtín ola* literally Martin of oil, idiomatically Oily Martin. The name refers, in the first place at least, to the corpulent nestling which, in former times, was taken for the sake of its body oil; the same motivation is seen in ALAMOOTIE, GOYLIR. The Stormy Petrel is sometimes regarded as a Swallow, hence a Gaelic name *áinleog mhara* literally swallow of sea. MARTIN is often used locally for the Swallow, and this usage has been imitated in Gaelic. Since the influence of English on Irish Gaelic is relatively recent, the present name can hardly have arisen before, say, the 17th cent.

Martin Snipe. A Norf. name for the Green Sandpiper, the conspicuous white rump of the flying bird recalling the familiar Martins. See SUMMER SNIPE.

Martlet. Unused today except as a term in heraldry, where it designates a Swallow-like bird, conceived as footless, the prototype being evidently the Swift. All the same, this appears to be secondary, the original here being French *merlette*, a diminutive of *merle* blackbird, cf. Cotgrave 1611 '*merlette* f. a martlet, in blason'. Otherwise the name has been applied to both the House Martin and the Swift. The earliest notice dates from 1536 'martlettes' glossing *Apedes* i.e. Swifts, the present spelling since Shakespeare (*Merchant of Venice*) 1596. The name is often quoted by lexicographers and naturalists down to Bewick 1804, surviving until the mid-19th cent. as a rare literary word; see MARTINET.

It has been supposed that Martlet in the sense House Martin or Swift arose from MARTNET, the contracted form of MARTINET, by substitution of the common final syllable -*let* for less usual -*net*. There is, however, French *martelet*, a variant of *martinet*; it is widespread in dialect, first recorded in 1560. There is no philological reason why this variant should not have developed in medieval times, in which case our English word could be, like Martinet itself, a direct borrowing from the Old French of the Normans.

Martnet, see MARTINET.

Mattock. A Ches. name for the Swift, being an Anglicization of the Welsh Christian name *Madog*, once a great favourite in Wales and also not uncommon in the the Welsh Marches. It is found in Ches. as a surname as early as 1290: Robert *Mattock* (Reaney 1976). The bird name will then be a local replacement of the more usual MARTIN, often found with the meaning Swift.

Mavis. A name for the Song Thrush, going back to c.1400, but now only in poetic or local use. It is an adoption of French *mauvis* recorded since c.1200: The earlier (pre-1150) form was **malvis*; this is shown by Spanish *malvis*, borrowed from French and reflecting the antecedent stage. The ulterior etymology is unknown.

Maw. A northern name for a Gull, chiefly Scottish, known since Holland c.1450. It is a loan from Old Norse *mār* (root *mā-*; see MEW), in records from Orkney and Shetland often written **Maa**. Since Ray 1678 'common Sea-Mall', the name is sometimes spelt **Mall**, a form without etymological justification, arrived at as follows. Since e.g. *caw* is the Scottish pronunciation of *call*, it was assumed that Maw should be properly written Mall, the true origin of the name not being perceived. See TIRMA, also FULMAR.

Mawp. A Lancs. name for the Bullfinch, corresponding to Dorset **Mwope**, originally **Mope*, the insertion of *w* being a local dialect feature (in the same way *most* has been pronounced *mwost*). These forms are variants of the widespread name NOPE. The apparently arbitrary change of consonants is paralleled in the formation of rhyming nicknames, e.g. *Bill* from *Will*; they may go back to medieval times, as in the case of *Ned* and *Ted* based on Edward, the records of which begin in the 14th cent. See MALPE.

Meadow Lark, Meadow Pipit, see PIPIT.

Mealy Redpoll. First in Gould 1837 'Mealy Redpole', apparently coined by him. The epithet refers to the mealy, i.e. greyish-white, parts of the plumage. The term won general acceptance, replacing GREATER REDPOLL. On the spelling, see REDPOLL.

Merganser, see RED-BREASTED MERGANSER.

Merle. Never an everyday word, but one proper to poetical language, denoting the Blackbird. It was borrowed from French, appearing first in 1450, coupled—as so often—with MAVIS: 'The Maviss and the Merle'. French *merle* derives from Latin *merula*; see OUZEL.

Merlin. First found *c*.1325 'merlyon', a form of the name continuing with variations until 1567 'Murleons'. Meanwhile a 15th-cent. source has 'merlyn' and this type asserts itself to become the norm from 1616 'Merlins'. It is regularly used by all the early naturalists.

The name presupposes Anglo-Norman *merillon*, with loss of the unstressed initial vowel from Old French *emerillon*, older *esmerillon*, a derivative of *esmeril*, a word taken up from Franconian German *smeril*. This German word, which survives locally as *Schmerl*, has a close parallel in Old Norse *smyrill*, implying a term of Germanic age. Nothing further is known about affinities, but the *l* suffix very probably indicates a diminutive formation, the Merlin being, of course, the smallest of the Hawks.

Mew. Although today a rare literary word, it is in fact the native English term for a Sea Gull, going back via Middle English *mēw* to Old English *mēw*, *mǣw*. It remained in ordinary literary use until the 17th cent. when it was largely ousted by the newcomer GULL. The name is Common Germanic, as seen by Dutch *meeuw*, German *Möwe*, further Old Norse *mār* (see MAW), most likely of imitative origin. It survived to a limited extent in local use down to the last century; cf. KIRR MEW.

Miggy. Used very locally in the north for the Magpie, a pet form comparable to MAG. It is relatively recent, more traditional northern names deriving from PIANNET.

Miller's Thumb, sometimes **Tom Thumb.** Local terms for various small birds: Goldcrest, Long-tailed Tit, Chiffchaff and related Warblers. Cf. THUMB BIRD.

Mire Bleater. A Northumb. name for the Snipe, the parts being borrowed from the more traditional names MIRE SNIPE and HEATHER BLEATER.

Mire Drum, Mire Drumble. Northern terms for the Bittern, the former an abbreviation of the latter. The primary source of the first one is Ray 1678 'Bittour or Bittern or Mire-drum'. The other goes back to 1398 (two spellings) 'mir drommel, myre dromble'. Turner 1544 has 'myredromble', Merrett 1667 'Mire Drumble'.

The word *mire* is of Scandinavian origin, hence the present terms will not have arisen before the Middle English period. Nevertheless, in a certain sense, they continue the earliest English name for this bird, seen in Old English *rāradumbla*, *rāredumle*. We assume that the first element (in Old English times felt as belonging to *rārian* to roar) eventually ceased to be understood and was replaced by *mire*, while the minor modifications in the second, onomatopoeic element reflect the general instability of such forms.

Typological considerations indicate that Old English *rāra-*, *rāre-* are corruptions of *rēara-* reed, lost in English but vicariously attested in the related German *Rohrdommel*, locally also *Rohr-dummel*, *-drummel* (*Rohr* reed). The second element is similarly unproblematic: we may postulate West Germanic *dumil-*, which consists of the agent suffix *-il-* added to the root *dum-* imitative of the booming call.

Mire Snipe. First found in Holland *c*.1450 'Myresnype', this local northern, especially Scottish, word for the Snipe was brought over by the Vikings; it goes back to Old Norse *mȳrisnīpa*.

Missel, etymological spelling **Mistle.** A name found in Pennant 1768, etc., an abbreviation of MISTLE BIRD or MISTLE THRUSH, apparently an oversight. The word is, in fact, the traditional term for mistletoe, now lost except in the above bird names; it goes back to Middle and Old English *mistel*.

Mistle Bird. A name for the Mistle Thrush first noticed in 1626 'Missel bird'. It was used by Ray 1678 'The Missel-bird, or Shrite', but was not taken up by later writers; see MISTLE THRUSH. It has, however, been recorded in living use in Shrops.

Mistle Thrush. First found *c*.1670 'miselthrush', in Ray 1678 (index) 'Missel-bird or Thrush'. Pennant 1768 has 'Missel', but his successors established the present name as the standard. See MISTLETOE THRUSH.

The bird is called after its fondness for mistletoe berries. But is the name traditional? The record is relatively late. The name is, however, paralleled in German *Misteldrossel*, though this, too, is late, not being found before the 16th cent. Now, throughout the history of English, THRUSH has commonly alternated with THROSTLE, but there is no trace of a term **Mistle Throstle*. These observations make it not improbable that the present term, and the German one also, are vernacular adaptations of Latin *Turdus viscivorus* mistletoe-eating thrush (*visci-* mistletoe). The origin of the synonym MISTLE BIRD is presumably to be sought in the same milieu. On the other hand we have, with the same motivation, the folk-name NORMAN GIZER.

Indubitably native names allude chiefly to the harsh cry, e.g. CHAR COCK, GAW THRUSH, HORSE THRUSH, SCREECH, SKIRL COCK, further SHRIKE 2, SHRITE.

Mistle Thrush names are sometimes applied, especially in Scotland, to the Fieldfare; see STORM COCK.

Mistletoe Thrush. A now obsolete name for the Mistle Thrush, first seen in Merrett 1667 'Mistletoe Thrush, or Saith' and occasionally found afterwards, the latest occurrence being apparently Fleming 1828. The term is a modernization of MISTLE THRUSH, which in the event has outlived it.

Mock Nightingale. A Norf. name for the Blackcap, a tribute to its vocal achievements. The name was reported by White 1770.

Moggy. A Sussex name for the Barn Owl, originally **Moggy Owl** or the like, the present word being a pet form of *Margaret*; cf. MADGE.

Moll Hern. A name for the Heron, widespread in the Midlands and further south, the nickname *Moll* denoting in the first place the female. The present term thus originally contrasted with JACK HERN.

Moll Washer, see WASHTAIL.

Molly Isaacs, see HAY SUCK.

Montagu's Harrier. A name given by Yarrell 1843, following Temminck 1820 'Busard de Montagu', in honour of the discoverer of this species. See ASH-COLOURED FALCON, ASH-COLOURED HARRIER.

Moor Blackbird, see FELL BLACKBIRD.

Moor Buzzard. A traditional name for the Marsh Harrier, making its debut in Ray 1678 'More-Buzzard', (index) 'Moor Buzzard'. *Moor* in the older language may have the meaning marsh; cf. MOORHEN. The name was adopted by Pennant 1768 and used by various writers up to Fleming 1842, and occasionally since. It survives as a local word in Cumb.

Moor Fowl. A common name in Scotland for the Red Grouse, often used in literature, the record going back to 1506.

Moorhen. Documented since *c*.1300 'mor-hen', the modern spelling first in 1655 'moor-hen', further Ray 1678 'Water-hen or More-hen'. As in MOOR BUZZARD, the word *moor* here preserves an old sense of marsh; cf. Merrett 1667 '*Gallus palustris* Moor-hen'. See WATER HEN.

Pennant 1776 introduced **Common Gallinule** as the name for this species and this was accepted by his successors as systematically appropriate. But Yarrell 1843 brought back the present name, since which time it has been the recognized standard.

Moor Peep. A Yorks. name for the Meadow Pipit—the bird is as much at home on a moor as in a meadow, and *Peep* may be compared to other echoic names for this bird; see further under PIPIT.

Mootie. A Norn name for the Stormy Petrel from Orkney and Shetland, an abbreviation of ALAMOOTIE.

Morr, see MURRE.

Morrel Hen. A name for the Great Skua from the Yorks. coast, *morrel* meaning dark brown, in reference to the dusky plumage. The peculiar use of *hen* suggests a noa name, cf. SEA HEN.

Moss Duck. A Scottish name for the Mallard, *moss* having the sense marsh, otherwise best known from place names, e.g. *Chat Moss.*

Moss Owl. A term from the West Yorks. fells for the Short-eared Owl. *Moss*, here literally meaning marsh, denotes the soft, marshy moorland, the bird's preferred habitat.

Mother Carey's Chicken. First noticed in 1767, this name will be a superstitious deformation of earlier, but unrecorded *Mother Mary's Chicken. It was undoubtedly a seaman's noa name, the bird being feared as a supposed harbinger of bad weather, see under PETREL.

Mountain Finch. This name is first attested in Ray 1678 (index) 'Mountain Finch or Brambling'. At first it competed with BRAMBLING. Halliwell 1847 uses it to define the local names FURZE CHIRPER, FURZE CHUCKER, but afterwards the term becomes less frequent. As the bookish word mountain at once indicates, the present name cannot be a folk-name. It is, in fact, a translation of Gesner 1555 *Montifringilla*, a word coined by Gaza 1476 from Latin *monti*-mountain and *fringilla* finch to render Greek *oróspizos* literally *oró*- mountain, *spízos* finch.

Mountain Linnet. A term for the Twite introduced by Ray 1678 'Mountain Linnet *Linaria montana'*, the English being a translation of the Latin. The name is appropriate since the breeding birds, at least, take to the hills and moorlands. This purely book name was commonly used by the naturalists of the first half of the 19th cent. and still regarded as the main name by Yarrell 1843 and subsequent editions; see further under TWITE. Cf. HILL LINTIE.

Mountain Ouzel. A literary name for the Ring Ouzel, introduced by Ray 1678 'Rock Ouzel, or Mountain Ouzel of Gesner, called . . . Berg-Amzel'. It is thus no more than a translation of Gesner's (local) German *Bergamsel*; it never took root in English and is no longer used. Native folk-names may show comparable motivation; see FELL BLACKBIRD, ROCK OUZEL.

Mountain Sparrow. A term introduced by Ray 1678 to name a species he had observed in Austria; it translates the Latin *Passer montanus* of Aldrovandi 1603. The bird was in fact the Tree Sparrow, which does not necessarily inhabit mountainous areas; nevertheless, the present name was used by several 18th-cent. authorities.

Moustached Tern, see WHISKERED TERN.

Mud Dauber, also **Mud Stopper.** Terms for the Nuthatch from various districts in the south. This species nests in holes in trees, lessening the opening by plastering the sides with mud.

Mud Plover. A local name for the Grey Plover alluding to the habitat: mud flats by the shore. Cf. SEA PLOVER.

Mud Stopper, see MUD DAUBER.

Muffie. A Scottish name for a Whitethroat, the 'little muff' being the distinctive throat band; also **Muffie Wren.**

Muffit. In Scotland a name for a Whitethroat, apparently an abbreviation of **Muffit** muff-like **Wren** (cf. MUF-FIE WREN) or a similar name. Also CHURR MUFFIT.

Muffler. An Oxon. term for the Great Crested Grebe. It alludes to the crest, *muffle* being a local expression for a tuft of feathers; see -ER.

Mullet Hawk. A folk name for the Osprey from Kent, Sussex, and Hants., from a supposed partiality to the grey mullet.

Mum Ruffin. A West Midlands name for the Long-tailed Tit. Its origin is undetermined. There are, however, similarities with other names for this bird: *mum* is reminiscent of *bum* in BUM BARREL, found in adjacent districts, while *ruffin* recalls ROSE MUFFIN. Such similarities imply a degree of interdependence and mutual contamination.

Murdering Bird. A Shrike name coined by Merrett 1667 '*Lanius* Butcher, or murdering Bird', a sort of rationalization of Turner 1544 'a shrike, a nyn murder'; see NINE MURDER.

Murdering Pie. A name for a Shrike, in local use in Worcs., Northants., and the north country. The first element is due to Merrett 1667 (see MURDERING BIRD) but the use of PIE in Shrike names is traditional; cf. FRENCH PIE.

Murre. First met with in 1602 in a list of bird names without identifications. It is reported in Ray 1678, who took down the name in West Cornwall in 1662, the meaning being 'Razorbill'. Also used for the (Common) Guillemot 1845, the sense recorded for the variant **Morr**

(Nance 1963). These words are relics of Cornish; cf. local Welsh *morr* Razorbill. In British usage, a local term confined to Cornwall and Devon, but in America Murre has become a common synonym for GUILLEMOT.

It is not unusual for the two species, Razorbill and Guillemot, to go under a single popular name, since they are superficially rather alike, and moreover consort. The Guillemot is, however, more numerous and also much noisier, the Razorbill being in fact a somewhat silent bird. An onomatopoeic name denoting both these species will therefore have arisen in reference to the Guillemot, as in the present case, the forms being imitative of the low, guttural call of the adult; cf. MARROT. Onomatopoeic names for this species are, in practice, more often based on the high-pitched, distinctive cry of the juvenile birds; see under GUILLEMOT.

Mussel Cracker, see OYSTERCATCHER.

Mussel Duck. A Norf. name for the Scaup, from the bird's chief food.

Mute Swan. The name coined by Pennant 1785 on discovering that his previous term TAME SWAN was inadequate as a specific term: 'We change the name of the Tame Swan into *Mute* . . . as this species emits no sound'. The epithet *mute* is, however, not strictly apposite, but the implied contrast with the strident Whooper is convincing enough and Pennant's revised usage was confirmed by his successors.

Mwope, see MAWP.

N

Naak. A name for the Great Northern Diver in use on the Northumb. and Berwick coasts, a representation of the loud call which commonly motivates Diver names.

Nan Piannet. A Yorks. name for the Magpie, tautological (see PIANNET) and apparently patterned on NAN PIE.

Nan Pie. A Yorks. name for the Magpie, reminiscent of PIE NANNY, but with reversed order of the two parts, presumably under the influence of the standard name MAGPIE.

Nanny Washtail, see WASHTAIL.

Nath. A North Cornwall name for the Puffin. Not recorded until 1836, but certainly traditional, being in fact a Celtic name, the otherwise lost correspondence of Welsh *nadd* something hewn or chipped off, a chip. The grotesque bill provided the motivation, as so often in Puffin names; cf. especially NORIE.

(Neb. The oldest word in our language for a bird's beak: Middle and Old English *nebb*, with a close correspondence in Dutch *neb*, Middle Low German *nebbe*. Today a northern word, it enters into certain local bird names, e.g. BLACK NEB, COULTER NEB. Cf. BEAK, BILL, and see under STINT.)

Nettle Creeper. A widespread local term for the Whitethroats, sometimes for the Blackcap or the Garden Warbler, species which commonly nest among coarse vegetation, such as nettles. An alternative, rather humorous name is **Nettle Monger.**

Nicker. A name for the Green Woodpecker, fairly common in the Midlands and north country. The dialect term *nicker* means a suppressed laugh and is also used of horses' neighing; it is itself an echoic word. The present name thus refers to the call, as quite often in the case of this bird; see YAFFINGALE. The relevant quotation from the *EDD* is instructive: 'Those nickers are calling out; they reckon it's a sign of rain'; cf. RAIN BIRD.

In JACK NICKER, the name has an entirely different origin.

Nickle. A name for the Green Woodpecker heard in Lincs. and Notts., a compromise between the synonymous ICKLE and NICKER.

Night Hawk. A fairly common term for the Nightjar found in the south country, so called from its nocturnal habits and the hawk-like appearance of the flying bird. Cf. DOR HAWK.

Night Heron. A species so little known in this country will never have had a genuine folk-name. Ray 1678 referred to Gesner 1555 *Nacht rab* i.e. Night Raven, from the bird giving voice during the hours of darkness. This German name apparently prompted Pennant 1785 'Night Heron', systematically more appropriate.

Nightingale. Old English *nihtegale* continues into Middle English, to be last recorded in 1483 'nyghtgale'. About the middle of the 13th cent., however, there appears a variant *nihtingale*, the source of the modern word. It appears that the (by then) unusual two-syllable *nihte-* of the Old English form was, in this case, altered to *nihtin(g)-* under the influence of a word-forming principle developing at the time, essentially as *railing* from *rail*, referred to under EASING SWALLOW. The older variant corresponds to Dutch *nachtegaal*, German *Nachtigall*, and is proper to West Germanic, originally **nachtagalō* literally night songstress; for a morphological parallel to **-galō*, cf. CROW. The word was traditionally feminine, our ancestors not realizing that only the male bird sings. On the other hand, they would for long retain a feeling for the literal meaning of this name, since the verb in question remained in use until late medieval times, cf. Old English *galan*, Middle English *galen* to sing.

Nightjar. So called from the unique jarring song, as with numerous other names for this species, including CHURN OWL, WHEEL BIRD. The record goes back to 1630. The early naturalists, however, generally preferred GOAT SUCKER, but Yarrell 1843 effectively standardized the present term. The Nightjar is, of course, active only after sunset.

Night Raven. A literary word used at all periods of the language. It goes back to Old English *nihthræfn*, a biblical word being a translation of Greco-Latin *nycticorax* occurring in Leviticus 11:16. It arose in the 7th cent., when the Bible began to be studied in this country. In the nature of the case, the name was little understood and vaguely identified with various birds in some way active at night, as the Bittern or the Nightjar.

Gesner 1555 applied *Nycticorax* to the Night Heron, in accordance with a use of the local German *Nachtrabe*. This meaning was borrowed into English by Ray 1678 'The lesser Ash-coloured Heron, or Night Raven'. It was adopted by Albin 1738, but in this strictly specific sense the present name was subsequently replaced by NIGHT HERON.

Night Reeler. The name in the Fens for Savi's Warbler, reported by Newton 1896. The bird nested in this area in some strength prior to the draining of its habitat in the first half of the last century. The name derives from the whirring song, similar to that of the Grasshopper Warbler, a bird popularly called REELER.

Nine Killer. A Shrike name, first in Ray 1678, a translation of Dutch *negendooder* or German *Neuntöter*, traditional terms for the Great Grey Shrike. The name derives from a superstition that the Shrike slays nine birds every day.

Through its occurrence in Ray, the word obtained a certain currency in bird books. Cf. NINE MURDER.

Nine Murder. Given as a Shrike name in Turner 1544 'a shrike, a nyn murder' and copied by a few later writers, last appearing in Cotgrave 1611 'ninmurder'. The entry in Turner is, however, a printing error, the name not being English at all, but German (modern spelling) *Neunmörder* literally nine murderer; cf. NINE KILLER. On German for English in Turner, see under SISKIN.

Nope. A name for the Bullfinch found in many parts of the country, the earliest record being Cotgrave 1611 'Nowpe', followed by Drayton 1622 'Nope'. It was a significant name in Ray's day, as shown by his entry in 1678 'Bulfinch, Alp or Nope', but he elsewhere characterized it as local or regional; see BULLFINCH.

The name has several variants. To begin with, Suff. **Ope**, from which would arise the variant **Hope** recorded from Somerset, Dorset, Devon, and Cornwall. It could be thought that Ope may derive from Nope as the result of false division after the indefinite article, i.e. *a nope* could be understood as *an ope*, just as *adder* arose from the original *nadder*. However, the opposite is equally possible, i.e. the presumptive original *an ope* could be felt as being *a nope*, comparable then to *newt* which developed from an earlier form without *n*; cf. Middle English *ewte*. As a third possibility, Nope could be envisaged as a rhyming nickname, standing in the same relationship to Ope as e.g. *Nelly* to its source *Ellen*, in which connection cf. the variants MAWP/MWOPE, POPE. On balance, it seems likely that Ope is the more original, since it can be plausibly matched with OLP, a variant of ALP, the oldest known name for the Bullfinch.

Norfolk Plover. Pennant 1768 classified the Stone Curlew with the Plovers. Knowing that it occurred in Norfolk (see STONE CURLEW), he bestowed upon it the present name which, however, he subsequently withdrew in favour of THICK-KNEED BUSTARD. Nevertheless, this purely book name has often been quoted by later writers.

Norie, recently also **Norrie.** A Shetland and Orkney name for the Puffin, first recorded in 1795, and to all appearances Norn; see -IE. Since Puffin names so very often allude to the singular beak, we refer the present word to Icelandic *nóri* something cut or chipped off, a chip (not directly recorded in Old Norse, but recognizable in *Nóri*, the name of a dwarf), and closely compare NATH.

Not surprisingly, the name has been popularly identified with the surname *Norie*, whence the form **Tammie** (Tommy) **Norie,** known since 1825, which together with a secondary **Tammie Norrie** (another form of the surname) is general in Scotland; also now Norrie in Shetland.

Norman Gizer, Norman Thrush. Names for the Mistle Thrush from Oxon. and Yorks. respectively. In both these localities, *Norman* has been recorded as denoting a tyrannical person: the allusion here will be to the aggressive behaviour of this bird towards other species during the breeding season. *Gizer* consists of the agent ending -ER added to *giz-* from Old French *guis* mistletoe; motivation then as in MISTLE THRUSH.

Norway Crow. A Yorks. and Norf. name for the migrant Hooded Crow. Cf. KENTISH CROW.

Nun. Formerly the best known traditional word for the Blue Tit, making its debut in Latin dress in Turner 1544 'nonnam'. The first occurrence in an English text is dated 1589 '*Parus minor,* a little titmouse, called a Nunne, because her heade is filletted as it were Nunlike' (*OED*). The name was still commonplace in the next century, witness Ray 1678 'The blue Titmouse, or Nun', but Pennant 1768 preferred the former, systematically more convenient term. The *EDD* records the present name

from widely separated parts of the country, while the *OED* has a citation from as late as 1903.

There is a French synonym, *mésange nonnette,* first recorded by Belon 1555, doubtless traditional. Perhaps the English term is ultimately due to French influence.

Nutcracker 1. Understandably this rare winter vagrant, *Nucifraga caryocatactes,* never acquired an English folk-name. The bird was described by Turner 1544, who knew it by one of its German names *Nussbrecher* literally nut breaker; this he translated into Latin as *nucifraga.* The same German name was known to Gesner 1555 who, however, chose to render it in Greek as *caryocatactes.* These Latin and Greek designations were available to Edwards, who in 1758 coined the English name; it has been used by all subsequent writers. The nut referred to is that of the Siberian cedar, the seed of *Pinus cembra,* a favourite food of this species.

Nut Cracker 2, see NUT JOBBER.

Nuthatch. The name is attested in three older variants: *c.*1340 'notehache', *Prompt.Parv. c.*1440 'nuthak', 1530 'Nothagge', the present spelling dating from 1668. Ray 1678 has 'Nuthatch, or Nutjobber', but Pennant 1768 used only the former, thus *de facto* establishing it as the standard. We trace the form back to a postulated Old English **hnuthæcca* literally nut hacker, the second element being the agent noun corresponding to the verb *hæccan* to hack. In the subsequent evolution of the language, this verb came to sound like the unrelated (Middle English) *hachen* to hatch (of eggs) and passed out of use. The variants 'nuthak' and 'Nothagge' (above) are parallel formations reflecting the still living verbs *hack* (from Old English *-haccian,* a variant of *hæccan,* as in the Woodpecker name WOOD HACK) and synonymous *hag* (northern, now only dialectal). The motivation of the name is not in doubt: nuts and acorns are wedged in crevices in the bark of trees and hacked open with the bill; cf. also NUT JOBBER.

For another local term, but with quite different motivation, see MUD STOPPER.

Nut Jobber. A name for the Nuthatch, first found in Turner 1544, and noted by Swainson 1885 as a Berks. word; the literal meaning is nut pecker, from the now obsolete verb *job* to peck, jab. In Northants., the name **Jobbin**, also spelt **Jobin**, represents a further development, after the analogy of ROBIN. Similar motivation is seen in the Shrops. name **Nut Cracker** and, of course, in the standard term NUTHATCH itself.

O

Oar Cock, see WAR COCK.

(-ock. A suffix traditionally used in the formation of bird names, going back via Middle English -*ok* to Old English -*uc*; see RUDDOCK. It remained productive, cf. TINNOCK, and reached Orkney and Shetland, see RITTOCK, RIPPOCK. In modern times, it has often been felt to be a diminutive, appearing in pet forms, when it may alternate with the more recent -IE, thus WRANNOCK beside WRANNIE.)

Olf, see ALP.

Olive. This chiefly East Anglian term for the Oystercatcher makes its debut in 1541/2 as 'Oliff', the regular older form of the (Christian) name; the present form dates from 1607. The synonymous North Frisian *liiw* and Dutch dialect *zeeliev* i.e. sea-*liev* show that the known English forms continue unrecorded Old English **lif*, secondarily assimilated to the Christian name. Here we have, therefore, our oldest name for the Oystercatcher, and to all appearances the traditional West Germanic name. It is clearly imitative, perhaps echoing the high-pitched alarm call; cf. CHALDER, TIRMA.

Olp, see ALP.

Oolet, see HOWLET.

Ope, see NOPE.

Oriole, specifically now **Golden Oriole.** The name Oriole is due to Pennant 1776; it is an Anglicization of Linnaeus 1758 *Oriolus*, a term going back to Albertus *c.*1250. This name, as Albertus himself states, derives from the voice; it is imitative of the most distinctive note, the clear, flute-like mating call of the male, represented by Old French *oriol* which lies behind Albertus' Latin. As an onomatopoeic name, it is reminiscent of the synonymous Modern German *Pirol*.

The term Golden Oriole first occurs in Buffon 1792, changing Oriole from a specific to a generic name, such usage afterwards becoming general. The epithet *golden* aptly describes the resplendent yellow body of the male; it ultimately reflects *aureolus* literally golden, a Latin form of the name, Albertus' term having been mistakenly so interpreted. Cf. GOLDEN THRUSH, also LORIOT. The native English name for this species is WOODWALL.

Orphean Warbler. This rarest of visitors was first found to be a British species in 1849. It was described by Yarrell 1856 under the name of 'Orpheus Warbler', introduced by Gould 1837 following Temminck 1820 who had coined (the now disused) *Sylvia orphea* as a tribute to the bird's vocal qualities. Subsequently the present form of the name became usual and appears in BOU 1883; the first record we find in *Ibis* 1868, p. 73. See GARDEN WARBLER.

Ortolan, also **Ortolan Bunting.** A passage migrant, too scarce to have acquired a native English name, the plump little bird became known in this country as an epicure's delicacy imported from France. The name first occurs in English in 1656, in French in 1552. However, the word comes from Provençal—the birds were netted in great numbers in the south-western region of France known as Landes—in which language it has the literal meaning gardener and goes back to Late Latin *hortulānus* deriving from *hortulus*, the diminutive form of *hortus* garden. The facile explanation that the bird is so called because it frequents gardens—how many other species do just that?—can hardly be the ultimate answer. The term as we have it must surely be a corruption, due to folk etymology, which has obscured the original shape of the word.

The Ortolan belongs to the Buntings, hence Linnaeus 1758 *Emberiza hortulana*. In imitation of this is the ornithologist's **Ortolan Bunting**, first noticed in Mudie 1834.

Osprey. First *c.*1460 'Ospray', the present spelling in Turner 1544. The word was borrowed with (unexplained) change of *f* to *p* from Old French **osfraie* (becoming *orfraie* 1491), itself from Latin *ossifraga* (also *ossifragus*) literally bone-breaker. Such a name can hardly be primary, but will be a folk-etymological corruption of an unknown, earlier form.

Certain 17th-cent. naturalists applied the name to the White-tailed Eagle (see SEA EAGLE), but since Pennant 1768 the name has been correctly identified with *Pandion haliaëtus*.

Otterling. A name for the Common Sandpiper, locality not given. The word has been formed by adding the suffix -LING to a verbal form *otter*, widespread in dialect and meaning totter, shake, also (and originally) *hotter*, making it most likely that the present name arose as **Hotterling*. The name thus alludes to the conspicuous head-bobbing and tail-jerking so characteristic of this bird.

Ouzel, also **Ousel.** The oldest English name for the Blackbird, reaching back through Middle English *ōsel* to Old English *ōsle*, this in turn by regular change from preliterary **amsle* postulated from the German cognate *Amsel*. It is ultimately related to Latin *merula* (see MERLE) presupposing **mesula* and, with expected loss of intervocalic *s*, to Welsh *mwyalch*, where *-ch* is a secondary suffix: the Old British form would be **misalkā*. A name with such an ancestry must go back into the second millenium BC. Its exact form at that early date eludes us, but the consonant sequence is clear: it is *m-s-l* and we assume the basic sense to be black, as is usual with Blackbird names in any language.

The present word remained in common use until the 17th cent., by which time BLACKBIRD was becoming the usual term, Ouzel itself continuing in literary language as an archaism. It survived better in certain dialects, sometimes as BLACK OUZEL. It lives on locally in BROOK OUZEL, WATER OUZEL, and in the standard nomenclature in RING OUZEL.

In Early Modern English, an initial *w* developed in some districts, and this form may be spelt **Woosell** as in Shakespeare (*Midsummer Night's Dream*) 1590 'The Woosell cocke, so blacke of hew, / With Orenge-tawny bill'.

Owl. This general term for any species of Owl goes back via Middle English *oule* to Old English *ūle*, etymologically identical with Dutch *uil*, Middle Low German *ūle*, also Old Norse *ugla* (in which language intrusive *g* is a regular development). These forms presuppose Proto-Germanic **uwwalō*; cf. further closely related German *Eule* presupposing **uwwilō*. These words appear to contain the diminutive element *l* (see -LE) attached to a root *uww-* echoing the Owl's hooting.

The present word is often used in folk-names for the Nightjar; see FERN OWL.

Owlard. An Owl name from Ches. and Shrops., from OWLET by suffix change; also HOWLARD, HOWLET.

Owlet. Records of this word reach back to 1534. Though commonly understood as denoting a small or nestling Owl, this diminutive form may occur locally instead of OWL, in which usage it overlaps with HOWLET. As to origins, the present word arose from Howlet under the influence of Owl.

Ox Bird. A name for the Dunlin, known especially from East Anglia and Kent, but also reported from Hants. and Lancs., the record going back to 1547/48 'oxburdes'. The name has not been explained, and one must reckon with the possibility that it is not to be taken literally, but rather as a corruption of an unknown antecedent form.

Ox Eye. A widespread name for a Tit, first encountered in Turner 1544 'great titmous or great oxei'. It is a translation of French *œil de bœuf*, a popular term for a small bird, commonly the Wren. The association of this term with Titmice is not surprising since the basic sense of TIT is small creature. The French expression will have become current here during the Norman period. It was widely used to denote any small thing; the expression *bull's eye* is simply another way of translating the same term; among bird names cf. BULLFINCH.

The sense of the present name has sometimes been extended to include other small birds, such as the Tree Creeper, witness 1589 'An Oxeye, or Creeper, *Certhia*'; see also HAWK'S EYE.

Ox-eye Creeper. A name for the Tree Creeper occurring in 17th-cent. sources, beginning with Merrett 1667.

Oystercatcher. Coined by Catesby 1731 to name the local American species, described as feeding on oysters. The brand-new term appeared in a learned publication in England the following year (*OED*) and was soon familiar in naturalist circles. Pennant 1768 speaks of the 'Sea Pie or Pied Oyster Catcher', but in 1773 uses 'Oyster Catcher' generically, reserving 'Sea Pie' for the native species. Succeeding ornithologists, however, mostly preferred the present, inelegant coinage, which was finally established by Yarrell 1843, at least as the main term. Since there is only one European species, it is generally unnecessary to qualify it. Needless to say, the name is inapposite, as the European bird does not 'catch' oysters; MUSSEL CRACKER of the Hartlepool fishermen is much nearer the mark. For the oldest English folk-name, see OLIVE.

P

Padge. A name for an Owl, particularly the Barn Owl, reported from Notts., Leics. and Warwicks.; abbreviated from **Padge Owl, Padge Owlet**, found in the same districts. Padge is an alteration of synonymous MADGE, both names being simply pet forms of the Christian name *Margaret*, the change comparable to *Polly* from *Molly*, pet forms of *Mary*.

Pairtrick, Paitrick, see PARTRIDGE.

Pandle Whew, see WHEW.

Parson, A nickname for the Cormorant along the south coast (Sussex, Hants.), doubtless an allusion to the sombre, clerical plumage and the typical stance, upright with wings outstretched.

Partridge. This name, the present spelling first in 1579, goes back to Middle English times: *c*.1290 'partrich' beside later recorded, but phonologically more original 'pertriche' *c*.1400. It came to us through the Normans, cf. Old French *pertriz*, the syllable *-riz* being changed in English to the familiar-sounding *-rich*. This French form is a variant of the more usual *perdriz*, with intrusive *r* beside older *perdiz* from Latin *perdix*, stem *perdīc-*; the Latin word itself was borrowed from Greek *pérdix*.

Middle English *-rich* appears in northern and Scottish texts as *-rik* (cf. *church* with Scots *kirk*), the earliest attestations being 1388 'pertrikis', 14th cent. 'partrykes', and these are the antecedents of local Scottish **Pertrick**, most commonly spelt **Pairtrick**, and northern **Partrick**.

From the 16th cent. on, spellings occur which show dissimilatory loss of the first *r*: 1578 'Patrich', whence today local **Patridge**. Parallel forms are found in Scotland and the north country, namely **Paitrick** and **Patrick** respectively, both widespread, the latter naturally now felt to be the Christian name.

The modern syllable *-ridge* from Middle English *-rich* is a regular develop-

ment in an unstressed final syllable. In dialect, however, the older form may occasionally survive, as very local **Partrich, Patrich**.

There is no trace of an earlier, native English name for this species.

Pea Bird, see PEE BIRD.

Peacock. The name first occurs as a personal name in Domesday Book 1086 'Pecoc', not actually appearing in reference to the bird itself until *a*.1377. More modern spellings, reflecting regular sound change, are 1553 'Peacockes', 1592 'peacock's'. Pennant 1768 introduced the specific use of the name. It is clearly a compound, Late Old English *coc* being added to *pē-*, earlier Old English *pēa*, an independent word deriving from Latin *pāvō*. The provenance of the Latin is obscure, but evidently onomatopoeia played at least some part in shaping the word, cf. *paupulāre* to cry (when speaking of a Peacock), further Greek *taôs*, also *taón*, which seem to be basically the same word, the differing initial consonant suggesting imitative variants. The bird itself was introduced to the classical world from India, so that some unknown oriental name may ultimately lie behind the Greek and Latin names.

Beside Middle English *pēcok*, there existed a contemporary *pōcok*, where *pō-* continues Old English *pāwa*, a more conservative borrowing of Latin *pāvō* than Old English *pēa* above. Although the form dropped out of use before modern times, it has survived as a surname, hence today *Pocock* as well as *Peacock*. The Old English form survives, too, as *Pow* or *Poe*.

Peahen. A name parallel to PEACOCK, the record going back to Middle English, the forms then being *pēhen, pōhen*.

Pee Bird. A Surrey name for the Wryneck, motivated by the loud song 'pee, pee'. Sometimes understood as **Pea Bird** and so written.

Peel Bird. A Sussex name for the Wryneck, basically imitative like PEE BIRD, but modified to match another synonym RIND BIRD.

Peerie Hawk. The Shetland term for the Merlin, *peerie* being the local word for little.

Peeseweep, see PEWIT.

Peet Lark, see TIT LARK.

Peeweep, Peeweet, Peewheet, Peewit, see PEWIT.

Peggy Dishwasher, see WASHTAIL.

Penguin. A name first in use among seamen to denote the Great Auk met with in Newfoundland. It has been known since 1578 'at one Island named Penguin . . . the birdes are also called Penguins, and cannot flie'; the name Island of Penguin possibly goes back to 1536 (*OED*). Edwards 1750 distinguished between this, the 'Northern Penguin', and the 'Southern Penguin' of the Straits of Magellan and the Cape of Good Hope. Pennant 1768 used both the present name as well as his own creation GREAT AUK, but regarding the latter as preferable, similarly Latham 1785 'Pinguin' beside 'Great auk'. Subsequent naturalists use the name only in the modern sense. The old meaning is, however, remembered in Newfoundland, and lives on there in the name of the *Penguin Islands*. But nothing has been discovered about its origin— Latham's spelling is due to mistaken speculation that the name was to be derived from Latin *pinguis* fat.
ZAA xvii 262–4 (1969)

Peregrine Falcon, or simply **Peregrine.** The ultimate prototype is the Latin *falcō peregrīnus*, literally foreign falcon, of Albertus *c*.1250, a term he applies to a young Peregrine which had been trapped when making its first long-distance flight from unknown parts—birds caught in this way were better developed than those taken from the nest. Albertus' Latin reflects vernacular usage seen in Old French *c*.1263 *faucon pelerin*, Old Italian 13th cent. *falcone pellegrino*, Middle High German *c*.1210 *bilgerin valke* (French *pelerin*, etc., from Late Latin *pelegrīnus* by dissimilation from Classical *peregrīnus*; German *bilgerin* borrowed from a Romance source). Eventually, this term for the more valuable birds came to designate the species as such, and this usage is found in Aldrovandi 1603. Ray 1678 adopted the name, easily Anglicized as 'Peregrine Falcon'; presumably with an eye on classification, he would feel it a better term than simply FALCON used by some earlier writers in this specific sense. Pennant 1768 followed Ray's precept, and his successors followed suit.

Especially during the present century, there has been a tendency to drop Falcon in this name, thus converting the original adjective Peregrine into a noun, though the full term has the backing of the official lists.

Interestingly enough, the above words had in some sort already entered the language before Ray 1678 began the modern tradition: first *c*.1386 'ffaucon peregryn', next 1456 'Fawken peregryne' and 1525 'Foucons peregrynes', the order of the parts reflecting the French source. From this 'Peregrine' (to denote the female, as often in falconer's parlance) was extracted, the quotation dating from 1575.

See BLUE HAWK.

Pertrick, see PARTRIDGE.

Petrel. First attested in this spelling in 1602, a corruption of 'pitteral' (i.e. *pitterel) and 'pittrel' which, though not actually recorded until 1676 and 1748 respectively, must nevertheless represent the original form. It was inspired by the jingle *pitter-patter* under the influence of the suffix -EREL, alluding of course to the birds tapping the water with their feet as they skim over the surface. The present corruption may have been originally no more than a misspelling; at any rate, the supposed connection with St. Peter, who walked on the waves, is certainly due to later speculation.

The specific **Stormy Petrel** was first taken up by Pennant 1776, in place of his own artificial 'Little Petrel' used in 1768. It had doubtless been created by seamen for whom the appearance of the bird was a portent of bad weather; see MOTHER CAREY'S CHICKEN, also WITCH. The use of the adjective *stormy* is in keeping with tradition, as explained under FIELDFARE, so that it is inappropriate to insist on a 'correction' to STORM PETREL, apparently initiated by Jenyns 1835. The present generic use of Petrel goes back to Edwards 1747 'Peteril'.

ZAA xvi 285–8 (1968)

Pettichaps, also **Pettychaps.** A name for the Garden Warbler, first reported by Ray 1674 'Pettichaps' from the Sheffield district; it is also known from Lancs. and Northants. This purely regional name was used by Ray 1678 'Pettychaps', thus gaining a place in ornithological literature. It is basically imitative of the cheerful song. Pennant 1768 took up the name, but uniting under it both the present bird and the Chiffchaff. Later, the two were sometimes distinguished as 'Greater and Lesser Pettychaps' respectively, until GARDEN WARBLER and CHIFFCHAFF emerged as the generally recognized standard terms. Even so, the present word was still occasionally quoted as an alternative name at the beginning of this century.

Pewit. The literary form of a group of onomatopoeic names for the Lapwing, dominant in the northern half of England and common in Lowland Scotland; they reproduce the penetrating, two-syllable call. The name has been known since c.1508 'puwyt', the present spelling first in 1725 and becoming usual. The name is basically **Peeweet**, heard locally, sometimes represented as **Pee-wheet**. A variant **Peeweep** may lead to **Peeseweep**, the latter particularly in Scotland. Although not attested in the medieval records—the north of the country is poorly documented in earlier literature—the present name seems to be traditional, cf. Dutch dialect *piewit*, German dialect *Piewitz*.

The forms described above may be closely compared to those arising from a related imitative term for this species; see TEUCHIT.

The present name is known from various districts as a term for the Black-headed Gull, first in Ray 1678 'Pewit or Black-cap', which Pennant 1768 made into **Pewit Gull**.

Phalarope. A name due to Brisson 1760, the French equivalent of the scientific *Phalaropus*, also coined by Brisson at the same time. The new word monstrously combines Greek *phalarís* coot and *poús* foot rendering 'coot-footed *Tringa*' (*Tringa* sandpiper), a name given by Edwards 1750. The exotic term was introduced into English by Pennant 1776, who described the two British species as 'Grey Phalarope' and 'Red Phalarope'. Needless to remark, neither of these birds is sufficiently in evidence to have acquired an English folk-name.

Pheasant. After the earliest attestation *a*.1299 'fesaund', the name appears in a medley of spellings which fall into two groups: those with initial *f*, known since c.1350 (see below), continuing until 1697 'Fesants', and those with *ph* from c.1390 'Phesant', the present spelling first in 1635, and general since 1750. In this word, final *t* is secondary (as in CORMORANT), the linguistically oldest form occurring c.1350 'fesauns'. The name comes from France; cf. Old French 1255 *fesan*, adapted from Latin *phāsiānus*, itself a borrowing of Greek *phasianós* (*órnis*) literally (bird) pertaining to *Phāsis*, the name of a river, now the Rhioni, which enters the Black Sea just south of Poti. It will be noticed that the spelling *ph* is a learned correction made under the influence of the classical sources.

Philomel. A poetic term for the Nightingale, which goes back to Greek *Philoméla*, properly the name given by Sophocles (5th cent. BC) to the mythical daughter of an Athenian king, said to have been changed into a Swallow or, according to another version, a Nightingale. The latter interpretation was adopted by Roman writers, from whose work the term found its way into medieval vernacular literatures. English references, however, starting *c*.1385, give the word as 'philomene', apparently a confusion with the saint's name *Philomena*. But Milton in 1634 innovates with 'philomel' in close imitation of the classical source, and this form at once set the standard.

The ancients evidently did not realize that only the male bird has the fine voice—nor did our own ancestors; see NIGHTINGALE.

Pianet, pronounced /paianet/, also with better spelling **Piannet**. Formerly the usual northern word for the Magpie and still widespread locally. The record goes back to 1594 'Piannet'. The name is quoted by Merrett 1667 'pyanet' and by Ray 1678 'Magpie, or Pianet'. It consists of *pi*- i.e. PIE pied and ANNET, originally *Annot*, a pet form of the Christian name *Agnes*, once exceptionally common in the north (Withycombe 1959). Cf. ANNET.

Pickatee. A Notts. name for the Blue Tit, echoing the bird's chirping. Cf. TINNOCK.

Pickerel. A Scottish name for the Dunlin or other small wading birds. It is formed with the suffix -EREL from *pick* to peck, grub, the motivation being the tapping and probing for worms and the like. For the sense, cf. TANGLE PICKER. See also DORBIE.

Pick Maw. A Scottish term for the Black-headed Gull, going back to 1450 'picmaw'. *Pick* is the northern form of

pitch, an allusion to the black head. Cf. PICK TARNIE.

Pick Tarnie. A Scottish name for a (Common or Arctic) Tern, first in 1784 'picterne'. *Pick* is *pitch* (as in PICK MAW) in reference to the black head-feathers, while TARNIE consists of *Tarn- (cf. TERN) with the diminutive suffix -IE.

Pie. The ultimate source of most English names for the Magpie. It is a borrowing from French, making its debut *c*.1250. French *pie* itself derives from Latin *pīca*. The present name was for long a normal literary term, thus Charlotte Brontë in 1853 'chattering like a pie' (*OED*), while Newton 1896 still uses it freely. In local speech, it maintained itself best in the north. It may enter into the formation of names for other species, as HORN PIE, SEA PIE.

The present word is used attributively as the first element in compound names of popular origin with the meaning pied, as PIE FINCH, etc., below. See MAGPIE, PIET, PIANNET, also HAGGISTER.

(Pied. An element in names of literary origin, genuine folk names having PIE but see PIEDIE.)

Pied Flycatcher. A term coined by Pennant 1776 and taken up by his successors, he having previously used COLD FINCH. This rare and retiring bird does not appear to have had a genuine, traditional English name. It was in the past confined to Wales, at least as a breeding species—it has several popular names in Welsh; see Parry 1962.

Piedie. A Ches. name for the Chaffinch, evidently formed from PIED with the suffix -IE; cf. PIE FINCH.

Pied Wagtail. Apparently first used by Latham 1783 for the species previously called WHITE WAGTAIL. This neologism may be said to have become usual since Selby 1825.

Pied Woodpecker. A term for the Great Spotted Woodpecker introduced by Bewick 1804, but not taken up by other naturalists, unlike his BARRED WOOD-PECKER, created at the same time.

Pie Finch. A West Midlands term for the Chaffinch, motivated by the pied plumage as in the synonymous PIEDIE; see PIE.

Pie Mag. A Yorks. name for the Magpie, the apparent reversal of the two parts being presumably due to the influence of synonymous PIE NANNY.

Pie Nanny. A Yorks. name for the Magpie, apparently arising by substitution of *Nanny* for ANNET, seen in the widespread synonym PIANNET. Since *Nanny* was becoming less usual after 1700 (Withycombe 1959), we assume the present term to have arisen before then.

Piet, pronounced /paiet/, older spelling Piot. A term for the Magpie, going back to Middle English *piot* which first occurs *c*.1225. It was known to Turner 1544 'piot' and Merrett 1667 'pyot', but is now a purely local, essentially northern term. It consists of PIE and a diminutive suffix; see -ET.

Pigeon. A word of French provenance, the earliest occurrences being *c*.1390 'peions', *c*.1430 'pyionys', *Prompt.Parv.* *c*.1440 'pyione', the present spelling first in 1483. The differences are, however, purely orthographic, reflecting similar variations in contemporary French sources, where the name has been known since the 13th cent., also with the earlier meaning of young bird, especially young pigeon. In English, too, the sense young pigeon is attested from *Prompt.Parv.* (above) until the early 17th cent. The name goes back via *pibiōnem* in the Late Latin of Gaul to Classical *pīpiōnem*, accusative of *pīpiō* young cheeping bird, cf. *pīpīre* to cheep, obviously of imitative origin. The phonetic changes in French, though considerable, are regular, cf. WIGEON.

An obsolete spelling 'Pidgeon' is occasionally found, as in Merrett 1667. Competing chiefly in the literary language with DOVE, and in regional usage with such terms as CULVER, CUSHAT, or QUEEST also, Pigeon acquired its present range, more or less, by the beginning of the last century.

Pigeon Felt. A name for the Fieldfare, familiar in Oxon., Bucks., and Berks. The allusion is to the pigeon-like blue-grey plumage; see further under FIELDFARE. Synonyms are BLUE FELT, GREY FELT.

Pigmy Curlew. A term for the Curlew Sandpiper introduced by Latham 1785 'Pygmy Curlew' and used by several early 19th-cent. writers.

Pigmy Sandpiper, see CURLEW SANDPIPER.

Pink. A local name for the Chaffinch, recorded from most parts of the country, purely onomatopoeic, from the reiterated, monotonous call. Similar patterning in the synonymous SPINK, also TINK, TWINK, CHINK. See FINCH, which in Scotland may be replaced by Pink, as in GOOLD PINK.

Pink-footed Goose. Identified scientifically and given its present name by Bartlett in 1839—in ignorance of Baillon's work in 1833, when it was declared to be specifically distinct from the Bean Goose. The names remain, though these are now considered to denote geographic races of a single species.

Pinnock. A Glos. name for the Hedge Sparrow, but to all appearances once more widespread, the record going back to *Owl & N. a*.1250 'pinnuc', later attestations including 1570 'Pinnocke, hedge sparrow'. The word is formed with the suffix -OCK from a syllable *pinn*, imitative of the piping note, with close parallels in HEDGE PICK, HEDGE PIP.

Pintail. First in Pennant 1768 as 'Pintail Duck' and to all appearances coined by him. *Pin* has here an older sense of pointed peg, a reference to the two elongated middle feathers of the tail.

Later writers may dispense with DUCK, and Pintail alone has been the recognized standard at least since BOU 1883. Genuine folk-names include SEA PHEASANT and CRACKER.

Piot, see PIET.

Pipit. For the early naturalists, as for ordinary folk, a Pipit was a sort of Lark, its commonest name being TIT LARK. Pennant 1768 drew attention to a previously unrecorded, minor name denoting 'a species (of Lark) taken in the neighbourhood of London and called by the birdcatchers a pippit'. Such a term is patently imitative of the call, as often with Pipit names, cf. PEET LARK, TITLING. Montagu 1802 used the present spelling in **Pipit Lark,** his term for the Meadow Pipit.

Bechstein 1795 separated the Pipits from the Larks to form a new genus *Anthus.* To match this with an equally distinctive English name, Selby 1833 seized on Pipit, at the same time coining names for the three British species which had been firmly identified by the beginning of the 19th cent.; these were 'Rock, or Shore Pipit, Meadow Pipit, or Tit, Tree Pipit', from which Yarrell 1843 selected the present standard **Rock, Meadow,** and **Tree Pipit,** expressly pronouncing against TIT LARK. **Rock Pipit** is patterned on Montagu 1802 ROCK LARK, also recorded as a traditional name. **Meadow Pipit** recalls Latham 1783 **Meadow Lark,** itself reflecting Merrett 1667 '*Alauda pratensis* tit-lark'. Finally, **Tree Pipit** is best explained in its creator's own words: 'Montagu very correctly observes, that this bird rarely alights upon the ground without previously perching on a tree, and that it also commences its flight from a tree, after leaving the ground' *op. cit.*

Plover. The record goes back to the early 14th cent., the name coming to us through the Normans, cf. Old French *plovier,* known since the 12th cent., the foreign ending -*ier* being replaced by the native -ER. The French word conti-

nues Late Latin **plovārius.* The name is basically imitative of the clear, far-reaching call, in this case heard as plō (the Norsemen heard it as lō, see SANDLO), but associated by folk etymology with *plovere* (Classical *pluere*) to rain, imparting a meaning something like rain bird. It is needless now to add that the many attempts to find a rational connection between the Plover and the rain, by writers ancient and modern, have necessarily been in vain.

A term of this sort would not, of course, be tied to a particular species; see also WHISTLING PLOVER. It has often been transferred to the Lapwing, cf. GREEN PLOVER, and 'Plover's eggs' are, of course, Lapwing's.

Plover's Page. A Scottish term for the Dunlin which, in the summer, may be seen in the company of the Golden Plover.

Poacher, see POKER.

Pochard. The name may be pronounced either /pocard/ or /poachard/, and has been so spelt. These differences reflect the twofold origin of the name as a development of the alternatives POKER and POACHER (see further under POKER), i.e. with substitution of -ARD for the original ending -ER. The earliest attestations are Turner 1544 'pochardae' a Latinized plural, next 1552 'pocharde' and 1598 'pocard', evidence that both pronunciations were current in the 16th cent., and implying, of course, that both Poker and Poacher must be older than these dates.

Johnson 1755 introduced the spelling 'Poachard', but Ray 1678 had had 'Poker, or Pochard', and Pennant 1768 took up the latter, which has been the standard since then.

Poke Bag. Recorded from Glos. and Shrops. as a term for the Long-tailed Tit. The expression is tautological, both *poke* and *bag* having the same meaning; they refer to the shape of the nest. Thus the name for the remarkable nest has been transferred to the bird itself, cf. BAG, FEATHER POKE.

Poke Pudding. An alternative name for POKE BAG Long-tailed Tit and found in the same districts. It is similarly tautological: pudding retains here its original sense of poke, bag, standing in the same relationship to *pud (see LONG POD) as e.g. *topping* to *top*.

Poker. An East Anglian name for the Pochard, first in Ray 1678 'Poker, or Pochard'. Such a term would originate as a nickname for various species of Duck, the meaning being eventually narrowed down to denote the present bird specifically. It derives from the manner of feeding, many Ducks characteristically poking their bills into this and that as they search for edible matter whether on land or in the water, the impression of poking being heightened by the movements of the flexible neck.

The verb *poke* has a collateral form *poach*, preserved in dialect, and having the same sense, being thus distinct from the better known homonym meaning to steal game. In America **Poacher** is used of the Wigeon. There can be little doubt that this variant of Poker was once known in this country too. Needless to say, the Wigeon is also a poaching i.e. poking Duck.

See further under POCHARD, where it is shown that both forms of the present name must be considerably older than the first attestation above.

Polly Washdish, see WASHTAIL.

Pomarine Skua, Pomatorhine Skua. The first of these was introduced by Stevens 1826, the epithet being adopted from the Latin of Temminck 1820 '*Stercorarius pomarinus*'. This was, however, an arbitrary version of *pomatorhinus* (ornithologist's Greek *pōmató*- lid, -(*r*)*rhīnos* nosed). The lid is the rim which develops over the base of the bill in summer, admittedly an inept term, seeing that all Skuas are equally pomatorhine. But it won through. Yarrell

1843 and subsequent editions took up the incorrect form, usually spelling it 'Pomerine', but BOU 1883 preferred the more accurate 'Pomatorhine', which also appears in Hartert 1912 and *Checklist* 1952. Nevertheless such official support has not sufficed to oust the corrupt competitor which still often occurs; after all, it is easier to say.

Pool Snipe, Pool Snite. Names for the Redshank, the latter from a source dated 1661, the former found in Ray 1678 'Redshank, or Pool-Snipe'.

Popard, see POPELER.

Pope 1. A Dorset name for the Bullfinch, apparently a rhyming nickname from *MOPE (see MWOPE); cf. *Polly* from *Molly* (originally *Mary*).

Pope 2. A West Cornwall name for the Puffin, first reported by Ray 1674. It can hardly be other than a relic of Cornish, in which language it will most likely have been a term denoting the grotesque beak, as in so many Puffin names, e.g. COULTER NEB, NATH, NORIE. Also called **Popey Duck**, obviously derivative.

Popeler. A Norfolk name for the Spoonbill, first occurring *c.*1400 'poplers', other early attestations being *Prompt.Parv. c.*1440 'Popelere *populus*', Paston 1459 'popelers'. Formed with the -ER suffix from **Popel**, twice recorded: *a.*1300 'popelles', 1597 'Poppel', with which compare, in addition to Latinized *populus* above, *poplorum* (genitive pl.) in a Latin document dated *c.*1300, further with suffix change **Popard** (see -ARD) present in a single citation of 1413 'popardys'. Affinities unknown, though one has the impression that a word of this shape will be of French origin.

Popey Duck, see POPE 2.

Popinjay. A local name for the Green Woodpecker, first in 1612 'poppiniaies or wood-peckers'. Properly an early term for a Parrot, gong back to the late 14th cent.; it is of exotic origin, coming into English through French. Parrots are commonly green and gaudy, which doubtless encouraged the present shift of meaning.

Post Bird. The Spotted Flycatcher frequently sits on top of a post to watch for passing insects, hence the present name, recorded from Kent and Sussex.

Pratincole. This species is met with here, as also in northern and central Europe generally, only as a vagrant and in these parts has never been prominent enough to acquire a popular name. In 1756, the Austrian naturalist Kramer coined the Latin term *Pratincola* (from *prātum* meadow, *incola* inhabitant) in reference to the bird's favoured habitat: meadows with short grass. Pennant 1773 then introduced this term into English as Pratincole.

Prine. An Essex term for the Godwit, with an early record 1547/48 'praynes'. The literal meaning of the name is awl, bodkin, an allusion to the long, slightly curved bill.

Proud Tailor. A Midlands term for the Goldfinch, here likened to a tailor proud of the clothes he makes, these being of course the splendid, many-hued feathers. The record goes back to 1770. Cf. SHERIFF'S MAN, also FOOL'S COAT.

Ptarmigan. First as 'Terminganis' in a Scottish text from the very end of the 16th cent., a rough representation of Scots Gaelic *tarmachan*, also *tarmach*, literally croaker, the hoarse croak of the bird being a familiar sound on the Highland moors. The name is made up of the imitative root *tarm* enlarged by the suffixal *-ach*, or *-ach* and *-an*, both common either singly or together, in Gaelic bird names; cf. SOLAN, TRILL-ACHAN.

The present spelling, with pseudo-Greek *pt-*, was introduced by Sibbald 1684. Pennant 1768 took up this form, which thus became *de facto* the standard.

Puck. A name for the Nightjar in Hants., Sussex, and Surrey, also **Puckeridge**, further in Sussex **Puck Bird**. The former words also denote a cattle disease, *puck* in this sense first in 1834, *puckeridge* occurring in White 1789 'The country people have a notion that the fern owl . . . which they call a *puckeridge* is very injurious to weanling calves, by inflicting, as it strikes, the fatal distemper, known to cow-leeches by the name of puckeridge.'

With no further evidence for these words, it seems impossible to reach a definitive etymology; see *OED*. We offer the following speculation. It is apparent that *puckeridge* could equally well be spelt *puckerage*, i.e. *pucker rage*, this then being the disease caused by the **Pucker* or Puck Bird, i.e. the bird that *pucks* strikes, of which Puck is an abbreviation. According to tradition, however, the Nightjar does not puck, it sucks; see GOAT SUCKER. Recalling that this mysterious bird of the night has been the object of superstitious dread (cf. LICH FOWL), one could interpret *puck* as an intentional deformation of *suck* in the service of name taboo, after the principle seen in MOTHER CAREY'S CHICKEN. It follows that the use of Puckeridge to denote the bird is secondary—if indeed the word was ever so used, for the style of White's statement suggests a slip in reporting.

Puffin, obsolete **Puffing.** Originally the name given to the cured carcass of the nestling Shearwater, until the end of the 18th cent. an esteemed delicacy, supplies coming from the Scillies and the Calf of Man. The word is first found in Latinized forms: 1237 'paphinus', 1297 'poffo' (stem 'poffon-'), the earliest vernacular attestations being 1490 'pophyns', 1502 'puffingis', Skelton c.1508 'puffin'. The nestling Shearwater is notoriously corpulent, and Puffing, Puffin simply mean fatling, the name being derived from the verb *puff* in the sense to be swollen, by means of the suffix -ING which became -IN in dialect. Such a word could not have developed in either Scilly or Man, since these places were at the time Celtic-speaking. It is more likely to have arisen in Anglo-Norman England in connection with the trade in the salted birds and, needless to say, it continued to denote the same subject as long as the trade flourished. Moreover, the term was also applied to the Shearwater by the early naturalists, as Willughby 1676 used the Latinized '*Puffinus*', still the scientific name.

In modern standard use, however, Puffin refers of course to an entirely different species, *Fratercula*. We envisage the change of meaning to have taken place as follows. Puffin, primarily a trade name, at length found its way to the Scillonian and Manx puffinries, and there came to denote the living Shearwater, eventually the adult as well as the juvenile, the same development seen in the synonym LYRIE. Now, both *Puffinus* and *Fratercula* nest in burrows in the same surroundings, and *Fratercula* too was of interest to fowlers. In these circumstances, some confusion of names resulted, as first noticed in Caius 1570 Latinized 'puphinus'. The new sense was reported in Ray 1678 'the bird called . . . Puffin in North Wales', distinct of course from his 'Puffin of the Isle of Man' in the same work; see SHEARWATER. Pennant 1768 used the term for *Fratercula* only, thus lending his authority to the new sense, which has since prevailed in correct ornithological writing. Nevertheless, it was not until the second half of the last century that the word in its new sense found more general acceptance. Up to that time, too, *Fratercula* was, as a rule, popularly known only by one or other of its local names, as COULTER NEB, NATH, SEA PARROT. See WHITE PUFFIN.
TPS 65–72 (1974)

Puffinet. A name for the Black Guillemot, first in Ray 1678 'the Puffinet of the Farn Islands', consisting of PUFFIN (*Fratercula*) with redundant suffix -ET—the Farne Islands name for the Puffin is COULTER NEB. The name of the Puffin is

otherwise occasionally applied to the present species, as Ulster **Black Puffin**. Martin 1698 equates Puffinet with SCRABER.

Puggy Wren. A Surrey name for the Wren, *puggy* meaning short and squat.

Purple Sandpiper. A self-evident term coined by Walcot 1789 to denote a rather rare species without a folk-name of its own. This very appropriate term at once commended itself to subsequent writers.

Purre 1. A name for the Dunlin, occurring at various places, chiefly in coastal areas, and known since Ray 1678 'At *Westchester* they call them *Purres*'. It imitates a characteristic, long, purring call.

The present name was understood by the early naturalists as denoting the 'Winter Dunlin', once thought to be a distinct species, and was so used by Pennant 1768 and later writers. After 1815, however, when this Dunlin and the 'Summer Dunlin' were shown to be simply phases of one species, Purre very quickly dropped out of literary use, as explained under DUNLIN.

Purre 2. Today a local Ulster term for a Tern, but formerly more widespread, as implied by the first attestation seen in Cotgrave 1611 '*Alouette de mer* . . . a Purre'. The name is echoic; cf. DIP PURL and see further under TERN.

Puttock. A name, known since c.1450, denoting either the Kite, as Charleton 1688 'Kite, or Puttock' or the Buzzard, as Ray 1678 'the common Buzzard or Puttock'. In popular usage the Kite and Buzzard often go by the same name; though differing in appearance, they are relatively close in size and habits, and both hurl themselves, in a sudden swoop, onto the small mammals on which they prey. According to Swainson 1885 the term was used in eastern counties and in the Midlands where, indeed, it occurs in literature as late as the 19th cent. We postulate Old English *putuc from the root *put-* with the suffix -*uc* common in bird names; see -OCK. The word is doubtless cognate with synonymous Old English *pyttel, pittel*, which passed into Middle English to be last recorded *c.*1450 'Pitill'; it survives corruptly in Dorset DUN PIDDLE and Wilts. DUN PICKLE, names for the Marsh Harrier, originally **Dun *Pittle**. Old English *pyttel* shows the more archaic form which goes back via preliterary Old English **pytil* to Proto-Germanic **putilaz*, the root *put-* this time taking the agent suffix -*ilaz*; see -LE. The root in question is recognizable again in Middle English *puten*, presupposing unrecorded Old English **putian* meaning to put, putt, push, thrust, hurl, the bird names evidently deriving from the latter, older senses. Thus both Puttock and ***Pittle** literally mean, in modern idiom, swooper.

Q

Quail. Though often hidden in the corn-fields or elsewhere in deep grass, the loud call of the male at once betrays the bird's presence. Not surprisingly, this call may lead to name giving. It is clearly reflected, for instance, in the local name QUICK-ME-DICK. The present name is also ultimately imitative, as follows.

In certain dialects of West Germanic speech, the syllable *kwak- formed the onomatopoeic basis, to which was added the *l* diminutive suffix (as in THROSTLE) giving *kwakilō. This name was used in the dialect of the Franco-nians who became masters of Gaul at the end of the 5th cent., and from them the term passed into French, there evolving regularly to become *quaille* by the time of the first recording in the 12th cent. This form was brought here by the Normans and had entered English by the 14th cent., first as 'quayle', the present spelling dating from 1684. In France the name further developed into modern *caille*; the original form is today best preserved in Dutch *kwakkel*. The oldest documentary attestation is 8th cent. Latinized *quacula*. The foreign word replaced the Old English synonyms *ersc henn* stubble-field hen and *edisc henn* field hen, the latter surviving in Middle English *eddish hen*.

Queece, Queest. In the West Midlands and the south from Berks. to Devon, Old English *cūsceote* and its variant *cūscote* wood pigeon (see CUSHAT) underwent an unexplained modifica-tion, *cū* being replaced by *cwī* and *eo* or *o* elided. These irregular forms make their appearance in the first half of the 15th cent., written 'quyshte' and 'quyste' respectively. The second of these survives in modern times as Queest, or with shortened vowel, as Shrops. Quist. The final consonant may be lost, hence Ches. Queece. In the Midlands, the root vowel may be diph-thongized, giving **Quoist, Quoice.**

Queet. A name in north-east Scotland for the Common Guillemot, a variant of COOT, which in Scotland has come to denote this bird. Cf. BELL KITE.

Queue, see SHOOIE.

Quick-Me-Dick. An Oxon. name for the Quail, echoing the well-known call, in which connection compare the imita-tive basis of QUAIL. For another inter-pretation of the sound, see WET-ME-LIP.

Quink. An obsolete Scottish name applied to a species of Goose not securely identified and first occurring in 1551 'claik, quink and rute'; see also citation under ROUTHEROCK. Neverthe-less, a Goose name of this shape can hardly be other than imitative.

Quist, Quoice, Quoist, see QUEECE.

R

Rail. A popular name for the Water Rail and Corncrake (or Land Rail) alike. In view of the many similarities between the two species, it is natural that both can go under the same name; indeed, there was once a superstition that Corncrakes turn into Water Rails for the duration of the winter. The early attestations are *c*.1450 'Rales', 1482 'Rayle', Skelton *c*.1508 'rayle', the first unambiguously specific record being Ray 1678 'Rail or Daker-hen', i.e. the Corncrake.

The name was introduced by the Normans, cf. French (Picard dialect) *raille*; the word is first recorded as Old French *raale*, whence the present-day *râle*, also loosely applied to both species. The name is one with *râle* death rattle, originally simply rattle, scraping, in reference to the voices by which these secretive birds commonly betray their presence.

Rain Bird, Rain Fowl. Local terms for the Green Woodpecker, whose cry was held to presage rain; see NICKER. The former variant, first seen in Turner 1544 'rayn byrde', is fairly widespread, while the latter, the original term (see FOWL), was still being used in the 18th cent.; it first appears in *Prompt.Parv.* *c*.1440 'Wodehake, or reyne fowle'. Cf. WET BIRD.

Rain Goose. A name for the Red-throated Diver reported from Shetland and Caithness, the loud, eerie voice believed to portend rain. A word uniquely from this area will be basically Norn, developing from Old Norse *regn* rain, *gās* goose.

Ralph. A nickname for the Raven from Ches. and Northants., evidently induced by the harsh, rasping call, but partly reminiscent of RAVEN, as the alliteration implies.

Rantock. Reported as an Orkney term for the Goosander, apparently originally *Rant enlarged by the secondary suffix -OCK. It may be compared with **Ranty-pipes**, preserved in a traditional rigmarole from the islands and regarded as meaning Gander. One presumes these words to be basically Norn, but neither can be interpreted etymologically.

Rat Goose, see ROOD GOOSE.

Rattle Thrush. A Yorks. name for the Mistle Thrush, motivated by the harsh, rattling cry.

Rattle Wing. A Norf. term for the Goldeneye, suggested by the rattling noise made by the rapidly beating wings as the bird takes off.

Raven. A venerable name going back to Old English *hræfn*, cognate with Dutch *raaf*, German *Rabe* from Old High German *hraban*, further Old Norse *hrafn*, the ancient forms pointing to a common ancestor in Proto-Germanic *hravnaz. In this reconstruction, *-n-* and *-az* are inflexional elements added to a root *hrav-*, originally *hraf-*. A root of this shape is explicable as the product of pre-Germanic i.e. Indo-European *krop-*, imitative of the bird's hoarse voice and close to the root *krep-* seen in Latin *crepere* to creak, crack. See the introduction, p. 9.

In Scotland, this name is often replaced by CORBIE.

Razorbill. A name first found in Merrett 1667 'Colymbus . . . Razor bill', next Ray 1674, also 1678 'Razor-bill', at the time in use on the north-west coast of England. In the circumstances referred to under AUK, the term was taken up by Pennant 1770, subsequent authorities concurring. The name is an appropriate characterization of the bird's beak, which is remarkably sharp at the edges.

Red-backed Shrike. A term coined by

Pennant 1776 and accepted by his successors.

Redbreast. A name for the Robin, first noticed *c*.1401 'redbreast'. It survives, of course, in ordinary use, often as ROBIN REDBREAST. It was used by Pennant 1768 and his successors; it is the name given in BOU 1883, even Hartert 1912, although it had largely yielded to ROBIN by the late 19th cent. Redbreast is, relatively speaking, a neologism. It might be thought that its late occurrence in the record could be deceptive, seeing that comparable compound names exist in closely related languages, as Dutch *roodborst* or German dialect *Rotbrust* (usually in diminutive forms *roodborstje*, *Rotbrüstchen*), which would imply a common, hence ancient origin. Other considerations, however, indicate that this is not so. Firstly, the second element is variable, witness standard German *Rotkehlchen* or dialectal *Rotkröpfel*, containing respectively the diminutives of *Kehle* throat and *Kropf* crop. Secondly, none of these names is recorded before 1544; for the earliest period, on the Continent as in England, only simplexes are found; see RUDDOCK. We conclude that the compound names are secondary, apparently not arising until the later Middle Ages or at the beginning of modern times. Thus Redbreast and its Continental namesakes are not related in a genetic sense, but are the products of independent, though parallel, developments.

Red-breasted Godwit, see BLACK-TAILED GODWIT.

Red-breasted Merganser. After Edwards 1747 had called this bird 'Red-breasted Goosander', Pennant 1776 coined the present term, which was accepted by later writers. In practice, MERGANSER alone often suffices; it was created by Gesner 1555 from Latin *merg*-diving, *änser* goose, first treated as an English word in 1752.

Red Cap. A name, chiefly northern, for

the Goldfinch, alluding somewhat inaccurately to the red face and forehead; the record goes back to 1785. Also a very local Yorks. term for the (Lesser) Redpoll.

Reddag. A Hebridean term for the Kittiwake, a misprint for *Riddag, occurring in Martin 1698. Such a word is naturally from Gaelic, in which language it may be spelt *riodag*. It is a borrowing from Old Norse *ryta (see RITTO), extended by the Gaelic diminutive suffix -*ag*, common in bird names and corresponding to English -OCK. A parallel form is present in RITTOCK.

Reddock. A Dorset name for the Robin (Swainson 1885), but doubtless considerably older, see ROBIN REDDOCK. It is a variant of the traditional RUDDOCK under the influence of the adjective *red*.

Red Game. A name for the Red Grouse, recorded by Ray 1674 'The Redgame: *Grygallus minor*'. Cf. BLACK GAME.

Red Godwit, see BLACK-TAILED GODWIT.

Red Grouse. A term coined by Pennant 1776 (index) after the model of BLACK GROUSE. It was adopted by most of his successors, after Yarrell 1843 becoming the unchallenged standard. For folknames, see GOR COCK, MOOR FOWL.

Red-headed Linnet. In lay practice, Redpolls pass muster as Linnets. Following this precedent, Ray 1678 constructed 'greater and lesser red-headed Linnet' for the Mealy and Lesser Redpoll respectively, and such names were used by Pennant and other 18th-cent. naturalists.

Red Hoop. A Bullfinch name from Dorset, with motivation as in BLOOD HOOP.

Red Kite. Coined by MacGillivray 1840 to distinguish the species known in Britain from the Black Kite of foreign parts.

Red-Legged Crow. An obsolete term for the Chough, coined by Pennant 1768. It was used by his successors in that century, but after Bewick 1804 had introduced CHOUGH for *Pyrrhocorax pyrrhocorax*, the present term was hardly ever used.

Red-legged Partridge. This species was imported into Britain about 1770, but had already been given the present name in ornithological literature by Ray 1678; cf. also Cotgrave 1611 *'Perdrix rouge . . .* the great red legd Partridge, the French Partridge'. See FRENCH PARTRIDGE.

Red-necked Grebe. Coined by Pennant 1785 who described a specimen received from Copenhagen.

Red-necked Phalarope. As a counterpart to his 'Grey Phalarope', Pennant 1776 named this bird 'Red Phalarope' after Edwards 1750 had called it 'red coot-footed *Tringa*'. It was Sowerby 1806 who introduced the present, more precise designation. It was taken up by Gould 1837 and has been general since Yarrell 1843.

Redpoll. First noticed in Albin 1738 'Red Pole', apparently a name from the vicinity of London. Later authorities may prefer the better spelling 'redpoll' i.e. red head (first in 1772), but not until this century can it be said to have prevailed generally. See RED-HEADED LINNET.

Redshank. Called after the long, red legs, first in 1525 'redshancks', later attestations including 1549 'redeschank', 1570 'Readshanke', Merrett 1667 '*Haemantopus* Red shanks'. But the present form, i.e. with *-shank* (singular) has been general since Ray 1678 'Redshank, or Pool-Snipe'. It was the term preferred by Pennant 1768 and has since been standard use. Cf. GREENSHANK.

Red Stare. A local Yorks. development of REDSTART, with substitution of STARE starling. There is also a weakened form **Redster**; cf. SHEPSTER.

Redstart. Literally Red Tail; see -START.

Although the record goes no further back than 1570 'Redstarte', the name is doubtless traditional, as shown by the synonymous Dutch *roodstaart* (today commonly in the diminutive form *roodstaartje*) and Low German *Rotsteert*; there are parallel forms in Scandinavian languages, as Danish *rødstjert* or Swedish *rödstjärt*.

The present name had been partly replaced by RED TAIL, but Ray 1678 used 'Redstart', Pennant 1768 followed suit, and in this way the archaic name was established as the standard term. Cf. FIRE TAIL.

Redster, see RED STARE.

Red Tail. A local term for the Redstart, found in various parts of the country. It was used by Turner 1544 'rede tayle', this being the first attestation. It is a modernization of REDSTART.

Red-throated Diver. Termed by Edwards 1747 'Red-throat Ducker or Loon' which doubtless suggested the present name, coined by Pennant 1776 and regularly used since. Popular names include BURRIAN, LOON, RAIN GOOSE.

Red Thrush. A local Midlands term for the Redwing.

Redwing. First noticed in 1668. Ray 1678 has 'Redwing, Swinepipe, or Windthrush', Pennant 1768 taking 'Redwing', which thus became de facto the standard term. To judge by the limited area in which its synonyms are found, it seems that the present name was, by the 17th cent., already the most widely used name. It is not possible to determine its age precisely, but since WING is a loan from Old Norse, the bird name can hardly have come into existence before the Middle English period. At the same time, it is not inconceivable that it reflects an unrecorded Old Norse *rauðwængja once used in these islands, the form being reconstructed from the bird's Norwegian name *raudvengja*.

Ree, see REEVE.

Reed Bunting. Introduced by Pennant 1776 for systematic reasons, replacing REED SPARROW.

Reedling. A term from the Fens for the Bearded Tit, doubtless traditional and appropriate for a species so attached to the reed-beds. A word of this nature may be supposed to have arisen in the Middle English period; see -LING.

Reed Pheasant. A rather facetious Fens name for the Bearded Tit, alluding particularly to the long tail.

Reed Sparrow. A traditional name, known since the 15th cent., for any sparrow-like bird inhabiting a reed-bed, i.e. the Reed Bunting and the Reed and Sedge Warblers, meanings still attested locally. It was used, in the first of the above senses, by the early naturalists from Turner 1544 'Rede Sparrow' to Pennant 1768, but yielded to REED BUNTING, coined by Pennant 1776.

Reed Warbler. First appearing in Pennant 1812, systematically preferable to REED WREN. Predictably, Fleming 1828 took up the present term, which with Yarrell 1843 was finally established as the ornithological standard. See MARSH WARBLER.

Reed Wren. A name for the Reed Warbler, occurring in various southern counties from Essex to Worcs. This folk-name was introduced into ornithological literature by Latham 1787 and found favour with certain later authorities, as Selby 1825, Jenyns 1835, Gould 1837, before being effaced by REED WARBLER.

Reel Bird or **Reeler.** Fens names for the Grasshopper Warbler, its song being compared to the whirring reel of the old spinning wheel.

Reeve. The female of the Ruff, first noted in 1634 '12 ruff and reeve 3 doz.' There is a well-attested variant **Ree**, first in 1512 'rey', later in that century 'ree' (Swann 1912); it survived into the 19th cent. We do not think that Reeve or Ree can have

originally denoted the female, since there seems to be nothing about her appearance or manner striking enough to motivate a unique name. Attention centres on the male, with his gorgeous ruff and aggressive behaviour in the mating season. These are features likely to lead to name-giving and we suppose that the present names refer to one of these two features. We take Ree to be the primary form and compare dialectal *ree* frenzied, seeing the motivation in the male's aggressiveness; the word is traditional, going back via Middle English *rēh* to Old English *hrēoh*. Such use of an adjective as a bird name implies a medieval origin, as in DUN or SWIFT. Given the above analysis, Reeve appears as a secondary form, perhaps arising through association with this word in a traditional sense of officer, governor, in allusion to the ornate plumage; compare in principle SHERIFF'S MAN. At some later date, when the original meaning of Ree and Reeve had been forgotten, a new term RUFF came into use for the male, so that the present terms became attached to the nondescript female.

(-rel, see -EREL.)

Richardson's Skua. A name for the Arctic Skua in honour of J. Richardson, the Arctic naturalist (1787–1865), introduced by Jenyns 1835 after Temminck 1820 'Stercorarius Richardsonii'. It was often taken up and used, sometimes as an alternative to ARCTIC SKUA, well into the present century; it is found in Yarrell 1843, etc. and was the term chosen by BOU 1883.

Ridlaik. An obsolete Scottish name for an unidentified species of Goose; it occurs in a text of 1578, see under ROUTHEROCK. Origin unknown.

Rind Bird, also **Rinding Bird.** Sussex terms for the Wryneck. *Rind* originally meant bark, the bird commonly feeding on insects found in the bark of trees. Meanwhile, *rind* has changed its sense and this, to all appearances, led to a further name, see PEEL BIRD.

It seems that the earlier sense of *rind* was preserved locally when used as a verb meaning to strip bark. As a consequence, the name was interpreted as the bird which sings in the rinding season (i.e. in the spring), hence the modified Rinding Bird.

Ring Bird. A term (locality unspecified) for the Reed Bunting, the *ring* being the white collar of the male.

Ring Blackbird. A Northumbrian name for the Ring Ouzel which, apart from its ring, looks for all the world like a Blackbird.

Ring Dove. This still common name for the Wood Pigeon, first noticed in 1538, alludes to the white patch on either side of the neck. It occurs in Merrett 1667 and Charleton 1668, and was used by Pennant 1768 and most of his successors down to the beginning of this century, likewise BOU 1883. But Hartert 1912 preferred WOOD PIGEON.

The present name is paralleled in Dutch *ringduif*, known since 1620, beside *ringelduif* going back to 10th-cent. *ringildūva*, comparable to German *Ringeltaube*. It is therefore not unthinkable that Ring Dove, in spite of the late attestation, is in fact an ancient name.

Ringed Plover. Coined by Pennant 1785 and adopted by most succeeding writers, thus becoming the standard term. The epithet *ringed* refers, of course, to the black collar; cf. the folk-name RING NECK. Other folk-names include DUL-WILLY, SAND LARK, STONE HATCH, TUL-LET.

Ring Neck. A Yorks. name for the Ringed Plover, *ring* denoting the black collar which so markedly contrasts with the surrounding white plumage.

Ring Ouzel. First recorded by Ray 1674, *ring* referring to the white gorget. The term was used in Ray 1678, similarly in Pennant 1768, since which time it has been the accepted standard.

Ring Tail. Known since 1538, this local name for the female Hen Harrier was motivated by the bars (rings) on the tail of the female, once thought to be a separate species. See BLUE HAWK.

Ring Whistle. A Yorks. name for the Ring Ouzel, suggested by the clear, whistling contact call. Contrast HILL CHACK.

Rippock. A Shetland name for the Kittiwake, evidently altered from **Rittock,** the form used in Orkney, the dental sound being traditional Norse.

Ritto. An Orkney name for the Kittiwake, from the local Norn and going back to Old Norse **ryta* (witnessed in Faroese and Icelandic). Like most Kittiwake names, it is of onomatopoeic origin, as befits this noisiest denizen of the cliff face. Further **Rittock,** with diminutive suffix -OCK; cf. RIPPOCK, also REDDAG.

Road Goose, see ROOD GOOSE.

Robin. Originally the Christian name *Robin*, a pet form of *Robert*, both coming to us from France with the Normans, but not recorded as a bird name until 1549. Since *Robin* is not found as a bird name in French, the present use will have arisen in England. To all appearances, the name was first applied as an alliterative epithet qualifying RED-BREAST, possibly also RUDDOCK. Later, it achieved independence, most likely under the influence of MARTIN It then competed with Redbreast, becoming by the late 19th cent. by far the commoner name, and must now be regarded as the standard, indubitably so in ornithological work, where it is more appropriate for terminological reasons, making possible the effortless creation of names for exotic species, as Blue Robin or Yellow Robin. See ROBIN REDBREAST, also ROBINET.

There is a long-standing folk tradition that the Robin is married to the Wren, hence the verselet 'The Robin and the Wren / Are God Almighty's Cock and Hen', once familiar throughout the

British Isles wherever English was spoken. This fancied association led, in the Shetland Islands, to a confusion of names, the Wren becoming known as Robin. Only one of the two names was required in popular use, for whereas the (Shetland) Wren is very common, the Robin is rare and hardly in evidence. English did not begin to establish itself in the Shetlands until the 18th cent., so that the islanders, for whom English was still an acquired foreign tongue, simply didn't know what was what.

Robinet. A term for the Robin found locally in Lancs., West Yorks., and adjoining parts of Derbys., first recorded in 1425 'robynet, redbrest' (*EDD*). It was evidently modelled on MARTINET. The use of ROBIN as a bird name appears later; see under this name.

Robin Redbreast. Noticed first in Holland *c*.1450 'Robyn Redbrest', next Skelton *c*.1508 in the same form; it is found in Ray 1678 'Robin-red-breast or Ruddock' and has, of course, remained in popular use to this day. See ROBIN, REDBREAST.

Robin Reddock. An apparently obsolete name, recorded in a Glos. glossary of 1639. It is to be compared with ROBIN RUDDOCK; at the same time, it provides the earliest attestation of REDDOCK.

Robin Ruck. A local northern name for the Robin, evidently a contraction of ROBIN RUDDOCK.

Robin Ruddock. A northern name for the Robin, first recorded *a*.1566, following the same pattern as in ROBIN REDBREAST.

Rock Dove. A name reflecting the rocky coastal habitat, known since 1655. The synonym ROCK PIGEON was formerly preferred, but after Montagu 1802 had given both terms, later writers generally used the present name, which accordingly became the standard. The early naturalists had difficulty in distinguishing this species from the Stock Dove, the systematic problem being complicated by the then uncertain position of the domestic pigeon. Not until Selby 1825 was it made unmistakably clear that Stock Dove and Rock Dove are distinct, so that the definitive specific sense of these names is really no older than that. Cf. BLUE DOVE.

Rock Lark. A Northumb. term for the Rock Pipit; it was used by Montagu 1802.

Rock Ouzel. A northern term for the Ring Ouzel, first in Ray 1674. Other Ring Ouzel names inspired by the rocky nature of the upland habitat are Yorks. **Crag Ouzel** and Devon and Derbys. (Peak District) **Tor Ouzel** (*tor* high rock, rocky peak). Cf. FELL BLACKBIRD.

Rock Pigeon. A now rarely employed name for the Rock Dove, first in Cotgrave 1611. It was used by Charleton 1668 and Pennant a century later, but after Montagu 1802 generally giving way to ROCK DOVE, though the *OED* quotes the present name from Darwin, *Origin of Species*, 1859.

Rock Pipit, see PIPIT.

Roller. Occurring only as a vagrant, this bird has apparently never acquired a genuine English name. The present term first appears in 1663 'a curious bird call'd Rollar Argentoratensis', then in Ray 1678 'Roller', whence it became part of our ornithological nomenclature. The name comes from Gesner 1555, who reports that this species was called Roller at Strasbourg (*Argentoratum*) and the name was onomatopoeic in origin. The word is thus proper to the German of that area—also independently recorded in 1554 (Suolahti 1909)—and is cognate with German dialect *rollern*, a verb imitative of sounds made in rapid succession and particularly applicable to the cries of this noisy bird.

Rood Goose. A Scottish name for the Brent Goose. The often quoted variants **Rat Goose** and **Road Goose** are echoes

of Ray 1678, who was shown a specimen of this bird in Yorkshire. A further
variant appears in the abbreviated form
Rout known from older Scottish
sources: 1551 'rute', 1578 and 1632 'rout'
(*OED*). The name is Scandinavian in
origin, cf. Old Norse *hroðgās*, evidently
a slip for **hrotgās*, as implied by Icelandic *hrotgæs*, also abbreviated *hrota*, and
Norwegian *ratgås*, with the literal sense
of snoring goose, an impressionistic
name based on the bird's call; it occurs
also in West Frisian *rotgoes* (see ROTCHE)
and Dutch *rotgans*. It will be noticed that
all the English forms show the corrupt-
ing influence of folk etymology.

Rook. Always the recognized English
name for the species, it goes back
through Middle English *rōk* to Old
English *hrōc*; the word is of Germanic
age, cf. Dutch *roek*, German (16th cent.)
Ruch from Old High German *hruoh*,
further Old Norse *hrōkr*, presupposing
Proto-Germanic **hrōkaz*. By invoking a
pre-Germanic base **krōg-* or **krāg-* (both
theoretically possible), the name is seen
to be an onomatopoeic creation. In the
first place, this basic form would apply
to both the Rook and the Crows; see the
introduction pp. 8–10.

Roseate Tern. This species, first identified in 1812 (see COMMON TERN), was
so named by Montagu 1813 from the
pink flush on the underparts in summer.

Rose-Coloured Pastor. An alternative
name for the Rose-coloured Starling.
Temminck 1820 established a genus
Pastor, calling the present species *P.
roseus*. The Latin *pāstor* shepherd suggested itself since this bird, like the
Common Starling, may be seen on a
sheep's back searching for ticks. Selby
1825 treated the word as though it were
English when he gave the bird the
present name; it has since been used
side by side with ROSE-COLOURED STAR-
LING, even though Temminck's genus
has long been discontinued.

Rose-coloured Starling. Though strik-

ingly beautiful, this bird is so rare a
visitor that it understandably never
received an English folk-name. It is the
Merula rosea of Ray 1713, whence Edwards 1747 called it 'Rose or Carnation
colour'd Ouzel' and more succinctly
Pennant 1776 'Rose-coloured Ousel'.
This latter name was used by Bewick
1797, but in 1821 he amended it, for
systematic reasons still valid today, to
Rose-coloured Starling. This new name
soon had to compete with an even
newer one, ROSE-COLOURED PASTOR, the
term found in BOU 1883, but Hartert
1912 and *Checklist* 1952 have the present
name.

Rose Muffin, a name for the Long-tailed
Tit, locality not quoted. The epithet *rose*
will naturally refer to the rose-coloured
parts of the plumage, but *muffin*,
though obviously a folk-etymological
corruption, is otherwise unexplained.
See MUM RUFFIN.

Rotche. A name for the Little Auk,
occurring in Martens 1694 'Rotges', a
West Frisian term for Brent Geese; see
ROOD GOOSE. Misapplied to the Little
Auk and provided with an English-type
singular **Rotge*, this foreign word
gained currency among seafarers in
northern waters, from whom it was
first taken down *c.*1809 in Shetland as
Rotche. Seamen from the Netherlands
were formerly frequent visitors to the
area in question, which adequately
accounts for the present borrowing, in
this respect comparable to MALLEMUCK.
Fleming 1828 introduced Common
Rotche; he was followed by Selby 1833
and Jenyns 1835, but other authorities
retained LITTLE AUK.

Rout, see ROOD GOOSE.

Routherock. A spurious name, first
found in 1806, purporting to be an
Orkney and Shetland term for the Brent
Goose. It is due to a misprint in Leslie
1578, where *six* types of Geese are
named 'Quink, Skilling, Claik, Rout-
hurrok, Ridlaik'. Here the fourth item

really consists of two names, Rout and Hurrok; see ROOD GOOSE, HURROCK.

Royston Crow. A name for the Hooded Crow, often quoted in bird books. It is first noted in Cotgrave 1611 'Corneille emmantelée, the Roiston Crow, or Winter Crow', next in 1617 'Royston Crowes'. It was used by Merrett 1667, likewise by Ray 1678 and Pennant 1768, but Pennant 1776 substituted HOODED CROW. The reference is to Royston (Herts.), where migrant birds appeared on the heath in winter, the Carrion Crow being the resident species. The same principle of name-giving figures in several synonyms, e.g. COATHAM CROW, SCREMESTON CROW. See also KENTISH CROW, NORWAY CROW.

Ruddock. Widespread in local use and until the 19th cent. still common in literature, this oldest surviving name for the Robin goes back via Middle English ruddok to Old English rudduc, the first attestation being datable to c.1000. The word is formed from the root rud(d)- red, ruddy; on the suffix, see -OCK. There is a cognate in the medieval German name rüdeling, where -ling is another suffix. Of interest here, too, is an Old English synonym rædda

literally red (bird), comparable to Old High German rōtilo (with the same literal meaning, -ilo being a suffix) which survives in Swiss German Rötele. These latter examples confirm that in the oldest period of the Germanic languages the names of the Robin were simplexes, not yet compounds of the sort seen in REDBREAST. Cf. REDDOCK.

Ruff. First recorded in 1634 '12 ruff and reeve 3 doz.'. The name evidently derives from the remarkable ruff or frill of feathers worn by the cock during the breeding season, and so properly applies to the male, as Ray 1678 states 'The Ruff, whose Female is called a Reeve'. Pennant 1768 used the term specifically, thus beginning the tradition of the two senses current today. See REEVE.

For other part for whole names, see under SHAG.

Runner. The Water Rail is a bird which generally keeps to the ground, so that it is most often observed as it runs from one piece of cover to the next, hence the present Yorks. name. Cf. JACK RUNNER, also BROOK RUNNER. Similar motivation is seen in the synonyms BILCOCK, DARCOCK, SKITTY.

S

Sad Cock, Sed Cock, Sedge Cock, Set Cock. Ches. names for the Mistle Thrush, the last also in Derbys. The differing first syllables are probably folk-etymological corruptions of an unknown original; one can only guess at some ultimate connection with SAITH, SYCOCK.

Saddle Back. A term for a Black-backed Gull recorded at various points along the coast, the saddle being the black back.

Saith. A term for the Mistle Thrush, found in Merrett 1667 'Mistletoe Thrush, or Saith'. Origin unknown. See SYCOCK, SAD COCK.

Sanderling. A name motivated by the sandy shores which are the preferred habitat of this common winter visitor. It represents an earlier, unrecorded *Sander (see -ER) with the further suffix -LING added under the influence of synonymous SANDLING. A word of this shape cannot be very old and may be envisaged as having come into existence, at the earliest, towards the end of the Middle Ages. First recorded in 1602, it figures in Ray 1678 'The Sanderling . . . also Curwillet . . .'; its adoption by Pennant 1768 effectively made it the standard name.

Sand Lark. A Scottish term applicable to both the Ringed Plover and the Common Sandpiper, often in the form **Sand Laverock**.

Sandling. A local name for the Sanderling, first recorded in 1684, and an example of the many formations with the suffix -LING, which became especially productive in the Middle English period, when the present name most likely came into being. See SANDERLING.

Sandlo. An Orkney word for the Ringed Plover, a relic of Norn from Old Norse *sandlō* literally sand plover, *lō* being onomatopoeic in origin; see LOO and compare PLOVER.

Sand Martin. First seen in Charleton 1668 'Sand, or Bank Marten', then Ray 1678 'Sand-Martin, or Shore-bird'. Pennant 1768 adopted the name, whence it became the standard term. It has clearly been motivated by the bird's preference for cliffs or steep river-banks with sandy soil, where the nesting tunnel may be conveniently excavated; it is presumably based on earlier SAND SWALLOW.

Sandpiper. This distinctive name, originally applied to the Common Sandpiper only, was of course inspired by the shrill, piping call. It was first recorded by Ray 1674 who described it as a Yorks. word. In Ray 1678 we read 'The lesser *Tringa* or Sandpiper'. Pennant 1768 took up the name in this traditional specific sense, but in 1773 converted it into a generic term to match Linnaeus 1758 *Tringa*, the present bird then being named **Common Sandpiper** for systematic clarity. See SUMMER SNIPE.

Sand Snipe, see STINT.

Sand Swallow. A local term for the Sand Martin, known since Bewick 1797, but almost certainly very much older. Indeed, since Martins were formerly called Swallows, it becomes very likely that the present term served as a model for SAND MARTIN.

Sandwich Tern. Coined by Latham 1785, commemorating the town of Sandwich (in Kent) from which place specimens had been sent to him.

Sandy Loo, see LOO.

Savi's Warbler. This species was identified ornithologically in the Fen district in 1840, though it was already known to the fenmen as NIGHT REELER. The present standard name was bestowed by Yarrell 1843 to honour Paolo Savi, the Italian naturalist who first described the bird in 1824.

Sawbill. A widespread term for the Goosander and the Merganser, also for the Smew, reference being naturally to the serrated edges of the bill common to members of the genus *Mergus*. Also **Sawbill Duck**.

Saw Sharpener. A name for the Great Tit, recorded from places as far apart as Worcs., Norf., and Roxburgh. It alludes to the grating, metallic notes heard from the male during the breeding season, The same motivation is found in other local names, as Staffs. **Saw Whetter**, Norf. **Sharp Saw**, also CARPENTER BIRD.

Scare Devil, see DEVIL 2.

Scarf. Used in Orkney and Shetland, and in the north of Scotland, for the Cormorant or the Shag, recorded since 1701, but indirectly witnessed since 1450, see SCART. The word survives from the former Norn language, continuing Old Norse *skarfr*, a word of Germanic age, regarded as imitative of the harsh, croaking sounds made by these birds.

Scart. A Scottish name for the Cormorant or the Shag, first found in 1513. It stems from an older form **Scarth** known since 1450, a differentiated variant of the more original SCARF.

Scaup. First reported as **Scaup Duck** in Ray 1678, similarly Pennant 1768, the epithet *scaup*, a northern form of *skalp*, meaning mussel-bed. The name was thus prompted by the favourite item on this bird's menu, as in one of its other names MUSSEL DUCK. The (nonsensical) abridgement to Scaup, used according to Newton 1896 by wild-fowlers, first appears in an article by Latham 1797 (*OED*). But it only became common in writing in this century, and since the 1920s has been the usual term, hence *Checklist* 1952 'Scaup'.

Sclavonian Grebe, see SLAVONIAN GREBE.

Scobby. A name for the Chaffinch found sporadically in the northern part of the country. Diminutive in form (see -IE) and apparently somehow connected with the synonymous Cumbrian **Scop**, also **Scoppy**.

Scooper. A name for the Avocet, first recorded by Charleton 1668 '*Avosetta*, the Scooper'. The motivation is evident: with its recurved beak the bird sweeps the shallows to scoop up small marine life. See further under AVOCET.

Scoot, also spelt **Scout**. A rather widespread, chiefly north-country and Scottish name for the Guillemot, imitative of the call of the juvenile bird; it has been known since 1596 'skout'. As often with Guillemot names, it is also used for the Razorbill. Cf. SKIDDAW, SKUTTOCK.

Scop, Scoppy, see SCOBBY.

Scorie. An Orkney and Shetland word for an immature Gull, first recorded in 1795. It is a relic of Norn, from Old Norse *skāri*, presumably of onomatopoeic origin, reinforced by the diminutive suffix -IE.

Scoter. Originally a Yorkshire term for the Common Scoter, first in Ray 1674 'The Scoter: *Anas Niger*', then in 1678 'The black Diver or Scoter'. The word was copied by Pennant 1768 and so became effectively the standard term. Nonetheless, it is spurious. It must be a scribal or printing error for *Sooter i.e. soot-coloured duck—one of its German names is *Russente* soot duck—the bird being not unnaturally named after the striking black plumage of the male, as in the local names BLACK DIVER, BLACK DUCK.

The present generic sense is due to Fleming 1828 who instituted for the Scoters a new genus *Oidemia* (now *Melanitta*).

Scout, see SCOOT.

Scoutie Allan. An Orkney and Shetland name for the Arctic Skua, first recorded in 1701, *scoutie* meaning shitty, from the belief that the bird lived on the excrement of other seabirds; see ALAN and cf. DIRTEN ALAN.

Scraber. A Shearwater name from the Hebrides, known since Martin 1698 who, however, quite inexplicably used it for the Black Guillemot. It is Gaelic *sgràbair* with redundant suffix *-air* (corresponding to English *-ER*) beside more original *sgràb*. The Gaelic, in turn, comes from Old Norse **skràp-* (witnessed vicariously by Faroese *skrápur*). The word literally means tough skin, an allusion to the scragginess of the adult Shearwater in contrast to the bloated nestling. See LYRE.

Scray. A Tern name, first in 1666 'Seaswallow, Scray', next in Ray 1678, who recognized its onomatopoeic nature, writing 'In other places . . . they are called *Scrays*, a name, I conceive, framed in imitation of their cry'.

Screamer, see SCREECH MARTIN.

Screech, also **Screech Thrush.** Widely occurring local names for the Mistle Thrush, recalling the penetrating alarm call. Similarly SHRIKE 2, SHRITE.

Screecher, see SCREECH MARTIN.

Screech Martin. Needless to say, the high-pitched call of the Swift has motivated this widespread popular name as well as many other comparable terms found in local use, among them **Screecher** and **Squeaker** occurring in various southern counties, further Sussex **Screamer** and Ches. **Squealer**; see also BLACK SCREECH, DEVIL SCREECH, then SCREW, and finally HORSE MARTIN—Swifts are often known locally as Martins.

Screech Owl. A widespread name for the Barn Owl, so called from the long, weird shriek, typical of this bird. The record goes back to 1593. Sometimes, less appropriately, identified with the

Tawny Owl, as in Merrett 1667 '*Strix* Screech, or screeching Owl' distinct from his '*Ulula* White hooping Owl'. The present name has now replaced an older variant **Scritch Owl**, recorded between Palsgrave 1530 'Scrytche houle' and 1813 'scritch-owl', and surviving in the southern United States (*OED Suppl.*).

Screech Thrush, see SCREECH.

Scremeston Crow. A Northumb. term for the Hooded Crow, from the number of migrant birds frequenting the seashore near Scremeston (Northumb.) in winter. Cf. COATHAM CROW, also ROYSTON CROW.

Screw. A Yorks. term for the Swift. It is imitative of the piercing cry, as if the bird were calling screw; we see comparable synonyms under SCREECH MARTIN.

Scribbling Lark, also **Scribbly Lark.** A widespread, chiefly northern name for the Yellowhammer, motivated by the flowing lines 'scribbled' over the surface of the egg. In Somerset, the name becomes **Scribbling Schoolmaster.** Cf. WRITING LARK.

Scritch Owl, see SCREECH OWL.

Scut. A name for the Wren in very local use in the south. The name literally means short, erect tail. Also **Scutty Wren,** sometimes abbreviated to **Scutty.** Cf. CUT.

Sea Cob, see COB.

Sea Dotterel. A name first appearing in Ray 1678 'Turn-stone, or Sea-Dotterel'; it was supplied by Sir Thomas Browne of Norwich.

Sea Eagle. A term introduced by Merrett 1667 'Sea Egle, *vel* Osprey' and copied by Ray 1678 'Sea-Eagle or Osprey'. The name is a translation of Greek *haliaetós* (*hali-* (salt) sea, *-aetós* eagle)—cf. Charleton 1668 '*Haliaetus, Aquila marina*, a Sea Eagle'—generally taken as denoting the Osprey, as in Pennant 1768. The name is found with this meaning until Selby

1825 used 'Cinereous Sea Eagle' for the White-tailed Eagle, which led Gould 1837 to use the present term for this species. In this sense it survives today, though it is not felt to be the ornithological standard.

Sea Gull. First noted in 1542, tautological for GULL.

Sea Hen. First in Ray 1678, this is the local name for the Common Guillemot along the north-east coast from the Tees to the Firth of Forth. It is a curiously imprecise term, hence the strong suspicion that it arose in fowler's parlance and was evasive in the first place. Guillemot eggs were often taken, which conceivably motivated the expression. Cf. TINKERSHIRE.

Sea Lark. A name widely applied locally to various birds found on the sea shore. It is first encountered in Merrett 1667 'Charadrios Sea Lark', i.e. the Ringed Plover, similarly Ray 1678. In more recent times, it has also been recorded as a name for the Dunlin, Rock Pipit, Sanderling, and Turnstone.

Sea Maw. First found in the 15th cent., tautological for MAW and used in the same general way, cf. synonymous SEA MEW. Ray 1678 spells the name **Sea Mall**, as explained under MAW.

Sea Mew. First in 1430, tautological for MEW and used in the same general way. Specific meanings—Common Gull, Herring Gull, Black-headed Gull— variously cited by certain early naturalists must have been arbitrary.

Sea Mouse. A name given to the Dunlin from Lancs. to Dumfries., elsewhere to the Sandpiper, motivation being presumably the scurrying to and fro typical of these birds.

Sea Parrot. A name for the Puffin, in local use especially along the east coast, obviously suggested by the appearance of the bill. The name is first attested in 1694. It cannot be very old, since *parrot* itself, evidently a neologism taken up

from some foreign source to denote a previously unknown bird, does not occur before c.1521.

Sea Pheasant. A local name for the Pintail, prompted by the long tail, and known since Ray 1678 'Sea-Pheasant or Cracker'. The expression is also applied to the Long-tailed Duck.

Sea Pie. A once well-known name for the Oystercatcher, first recorded in 1552, PIE prompted by the Magpie-like black and white plumage. For Pennant 1768 it was the obvious choice, but in 1773 he introduced the book word OYSTER-CATCHER. Nevertheless, the traditional name remained in common literary use until this century. Seeing that *Pie* is of Norman origin, the present term will hardly antedate the second half of the 13th cent., when Norman words first entered the English language in large numbers; perhaps it owes something to synonymous French *pie de mer*, too. For the oldest English name for this species, see OLIVE.

Sea Plover. The Grey Plover is locally so called since it keeps to the shore, where it feeds on small marine life. Cf. MUD PLOVER.

Sea Snipe, see STINT.

Sea Swallow. A name for a Tern, used by certain early naturalists, notably Ray 1678, who describes it as both an apt and common name. We doubt, however, if it is indigenous. It is first recorded by H. Hexham, *English-Dutch Dictionary*, 1647 'A sea-swallow *Een zee-swaluwe*', similarly German *Seeschwalbe* known since the 15th cent. Ray 1678 refers to 'Aldrovandus his Sea Swallow', also suggesting a term of alien provenance. (The concept Sea Swallow is well—and we think, more aptly—attested in Gaelic for the Stormy Petrel, see MARTIN OIL.)

Sea Turtle. An obsolete name for the Black Guillemot. The great attachment of the sexes, very evident in the complicated displays, led seamen to compare the bird with the Turtle Dove, hence the present name, recorded by Ray 1678 'Greenland-Dove or Sea-Turtle'.

Seave Cap. A name from Thirsk (Yorks.) for the Reed Bunting—*seave* (from Old Norse *sef*) is a widespread, mainly northern word for reed. The present, unique record seems to be a cross between *Seave Bird and BLACKCAP.

Sed Cock, see SAD COCK.

Sedge Bird. A term from Surrey and Sussex, first found in Albin 1738 and taken up in Pennant 1770, but in 1776 (index) replaced for systematic reasons by Sedge Warbler. The epithet *sedge* refers, of course, to the habitat, the sedge-fringed banks of streams and pools.

Sedge Cock, see SAD COCK.

Sedge Warbler, see SEDGE BIRD.

Serin. Originally a bird native to the Mediterranean area, the Serin has for some centuries been extending its territory northwards, and has latterly reached the south of England and bred there (1967). Serin is no more than the French name adopted for a bird which obviously had no native English name. It is first noticed in Ray 1678 (index). In France, the name is recorded as far back as 1478; it is known to have come into French from Provençal, but no etymology can be given. The word was well known to naturalists in its Latinized garb *Serinus* found in Gesner 1555, the form which Ray used in his text.

Set Cock, see SAD COCK.

Seven Whistler. A name for the Whimbrel, known from various parts of the country, its twittering, whistling note being imagined as repeated seven times.

Shad Bird. A Shrops. name for the Common Sandpiper. Until prevented by the building of weirs, the Shad used to ascend the Severn to spawning beds in the Shropshire area, an event coin-

ciding with the return of the Sandpiper to the same waters.

Shag. First recorded in 1556 'schagge', next in 1602 'Shags'. The term means what it says, i.e. tuft, and refers to the crest which develops during the breeding season. It is thus a name based on the principle of part for whole, as BEAK, RUFF, also AUK, SNIPE, further LAPWING. In popular use, this bird and the Cormorant may go under one name (see SCARF), but since the latter has no crest, the present term suitably distinguishes the species and, we suppose, arose for that express purpose, just as several names for the Cormorant refer to the unique seasonal white spot appearing on the thigh of that bird (see LAIRBLADE).

The earliest records refer to Cornwall, so that the name may have come into being in the eastern, i.e. English-speaking, parts of that county. It was used by the pioneer ornithologists and may be said to have been, from the beginning, the standard term.

Shalder. The Shetland name for the Oystercatcher, a Norn term from Old Norse *tjaldr*; see further under its Orkney equivalent CHALDER.

Sharp Saw, see SAW SHARPENER.

Shearwater. Originally a Norf. term for the Manx Shearwater, first in Ray 1674. In Ray 1678 the species was termed 'The Puffin of the Isle of Man', in the index 'Manks Puffin'. Pennant 1768 put 'Manks Petrel', but in 1776 took up the present term. Later ornithologists followed him, though in less careful style, and, in popular use, the bird was for long referred to as PUFFIN, a practice which has often led to problems of identification; see BOUGER, BULKER. The present generic use is due to Selby 1833.

The name reflects the unique flight pattern of the Shearwaters, the birds gliding over the waves on rigid wings, tilting over almost at right angles, now to this side, now to that, the tip of the lower wing seeming almost to cut the water. The same motivation is seen in the synonyms CLEAVER, HACKBOLT, see also COCKATHODON.

Sheel Apple, Sheely, see SHELD APPLE.

Sheep Rack. A Northants. name for the Starling, where *rack* means tormentor— the birds may perch on the backs of sheep and peck at the ticks living in the fleece. Cf. SHEPSTER.

Sheld Apple. A name for the Chaffinch, first in Turner 1544 *'Fringilla* . . . a chaffinche, a sheld appel, a spink', apparently once widespread, but in more recent times confined to the north country with isolated attestations in Northants. and Staffs. It was commonly altered by folk etymology to SHELL APPLE, when it occasionally denoted a different bird; see under this form. It also appears as *Sheel Apple* (to *sheel* being a dialect form of *to shell*). Pet forms **Shelly, Sheely,** are found, too.

The word *sheld* means variegated and is a traditional epithet, cf. the local German synonym *Schildfink* (-*fink* finch); it occurs again in our SHELDUCK. But *apple* remains intractable, though it has been conjectured to be a corruption of ALP. Cf. SHILFA.

Shelduck. Properly **Sheld Duck**, and often so written, the epithet meaning variegated as befits a species distinguished above all by the boldly contrasting colours of the plumage, common to both sexes. The term goes back to 1707, when it was given as **Shell Duck**, a form occasionally used since, and due naturally to a fancied connection with shells. See SKELDUCK, SKEEL-DUCK, SHIELDUCK.

The present name was preceded by **Sheldrake**, first attested as a bird name *c*.1325 'Sheldedrake', but found much earlier as a personal name: 1195 'Roger Scheldrac' and 1275 'Adam le Sceldrake', from Essex and Suff. respectively (Reaney 1976). There are variants of *shel-* corresponding to those associated with Shelduck, the oldest from 1600 'skeldraikis'. The name continued

in common use until the turn of the last century, by which time Shelduck was becoming general, cf. BOU 1883 'Common Sheldrake', but Hartert 1912 'Sheld-Duck'. The reason for the exceptional use of DRAKE as the generic part of a Duck name is given under BERGAN-DER.

Shell Apple. The cone-eating Crossbill is an uncommon visitor to this country and consequently without a regular folk-name. Nevertheless the bird may become prominent temporarily if, on the occasion of one of its rare eruptions, it suddenly appears in orchards, tearing at apples to reach the pips. Hence the present name, first noticed in Merrett 1667 'Shell-Apple'.

On reflection, however, the name seems rather inappropriate, since the Crossbills don't shell the apples, but in fact destroy them. The term can therefore hardly be an original Crossbill name, and we realize that it is simply an *ad hoc* application of a name properly denoting the Chaffinch, see SHELD APPLE.

Shelly, see SHELD APPLE.

Shepstare. A term for the Starling occurring in the northern half of the country; it is first attested in 1563, later recordings being 1584 'Stare or Shepstare' and 1681 'shepstares, stares or starlings'. In more recent times a reduced form **Shepster** is usual, sometimes altered, arbitrarily it seems, to **Chepster**; an abbreviated **Shep** and its pet form **Sheppie** are also current. Folk etymology is evident in local Yorks. **Ship Starling**.

Shepstare presupposes Middle English **shēpstare* literally sheep stare (starling) with subsequent regular shortening of the long vowel (as in *shepherd*). The bird has been so called from a habit of perching on the backs of sheep to feed on the ticks; cf. SHEEP RACK. See STARE.

Sher Cock, see CHAR COCK.

Sheriff's Man. A Shrops. name for the Goldfinch in a local source dated 1796, the ornate plumage being compared to the black and yellow livery of the sheriff's man. Cf. PROUD TAILOR, also FOOL'S COAT.

Shielduck. First in Pennant 1770 and occasionally used by others, apparently a compromise between SHELDUCK and Scottish SKIELDUCK.

Shilfa. A Scottish name for the Chaffinch, occurring in a medley of local variants, which appear to be allied to an old form **Shoulfall** known from Sibbald 1684. It is supposed that the second syllable -*fa*, -*fall*, could be a corruption of FOWL bird, the first element then vaguely recalling the English synonym SHELD APPLE.

Ship Starling, see SHEPSTARE. ·

Sholard, see SHOVELARD.

Shooie, also **Shoog.** A Shetland term for the Arctic Skua, clearly a word from the lost Norn language. Our starting point will then be the Old Norse name for this species **kjōi* (attested by proxy in Faroese and Icelandic). This word developed in older Norn first into **kjū*, but was at some point changed to **tjū* under the influence of **tjūv* thief (from Old Norse *þjófr*), which in the course of phonological evolution gave later Norn **shoo*, whence with the addition of the Scottish diminutive ending -IE the present Shooie. The variant Shoog, on the other hand, is purely Norn, excrescent *g* being an occasional feature (cf. Faroese *kjógvi* beside *kjói*); see also LOOG under LOO. Confusion with the word for thief requires no justification in the case of this robber bird, and developments comparable in this respect have taken place locally in the closely related Faroese and Norwegian. For another example of the regular change in Shetland Norn of *tj* to *sh*, see SHALDER.

Old Norn **kjū*, postulated above, has to all appearances survived in **Queue**, the Fair Island name for this species. Lockwood 1961, pp. 38–41.

Shore Bird. A name for the Sand Martin, reported by Ray 1678 'Sand Martin, or Shore-bird'. Such a name will presumably have arisen in some coastal district where suitable nesting sites were available.

Short-eared Owl. Coined by Pennant 1768 and generally accepted; the term contrasts, of course, with LONG-EARED OWL.

Shoulfall, see SHILFA.

Shoveland. A Norf. term for the Spoonbill, a variant of SHOVELER (see -ARD), first occurring in *Prompt.Parv. c.*1440 'schovelerd', next 1533–4 'shovelard'. In Sussex the form is **Sholard**, the word *shovel* appearing as *shole* in various dialects.

Shoveler, also spelt **Shoveller.** First found *c.*1460 'shovelere'. The name originally denoted the Spoonbill (cf. SHOVELARD), but Ray 1674 books the name as meaning *Anas clypeata*, also adopting this meaning in 1678. It was so used by Pennant 1768 and his successors, whence it became the standard sense.

Shrieker. An East Anglian name for a Godwit, motivated by the voice of these noisy birds.

Shrike 1. A name found first in Turner 1544 'a shrike, a nyn murder', said to denote the Great Grey Shrike. Turner declares that he had the word from Sir Frances Lovell, the only person he knew who could give him this bird's name; on 'nyn murder', see NINE MURDER.

A word of this shape strongly suggests onomatopoeia and a meaning of shrieker, but this is hardly applicable to a Shrike, indeed we recall no example, in any language, of a Shrike name deriving from the voice. We therefore follow Newton 1896 in regarding the name as properly belonging to some other species, perhaps the MISTLE THRUSH; see SHRIKE 2.

'Butcher Bird' was used by Pennant

1768, but in later editions he preferred the present word, and distinguished the two better known species as 'Shrike (or) Great Shrike' and 'Red-backed Shrike', thus confirming the generic use found in Ray 1678 'Butcher-Birds or Shrikes'.

Shrike 2. A local name in the south, but also in Cumb., for the Mistle Thrush, imitating the alarm note; cf. SHRITE, also SCREECH.

Shrill Cock. A Ches. name for the Mistle Thrush, motivated by the cry, and not far removed from its Derbys. equivalent SKIRL COCK.

Shrite. A north country name for the Mistle Thrush, first in 1668 and quoted by Ray 1678 'Missel-bird or Shrite'. The name echoes the scraping alarm note; cf. SHRIKE 2, also SCREECH.

Shuffle Wing. A name for the Hedge Sparrow recorded from places as far apart as Surrey, Worcs., and Yorks., motivated by a peculiar shake or flutter of the wings.

Silk Tail. A term for the Waxwing, first occurring in 1685 and occasionally used in the 18th cent. It is a translation of *Seidenschwanz*, the most widespread German name for this species, known since 1552. The name alludes to the soft plumage and provided the model for the scientific *Bombycilla* (Greek stem *bombyk-* silk, ornithological Latin *cilla* tail).

Siskin. First in Turner 1544 'siskin *anglice*'. The name was used by various later writers, including Ray 1678, and finally Pennant 1768 gave it his blessing in preference to BARLEY BIRD or ABERDEVINE, since which time it has been standard.

The word corresponds to Modern Low German *Siesken*, a form which will have been preceded in Middle Low German by *sīse*, *sīsek*, borrowed from Middle High German (13th cent.) *zīse*, *zīsik* (whence Standard German *Zeisig*). In their turn the High German words

came, through the trade in cage birds, from a Slavonic language, cf. Czech *čiž* and its diminutive *čižek*, of onomatopoeic origin. One could imagine that the name came into our language through similar channels, had not Turner expressly stated that the English called these birds 'aves canarias'.

Turner's book, written in Latin and produced in Cologne, gives side by side the vernacular English and German equivalents of the classical names of the birds described. And here lies the explanation of 'siskin *anglice*'; it is a mistake for 'siskin *germanice*', either through oversight or as the result of a printing error, as also seen in the cases of GEIR, NINE MURDER, STEINGALL, and WITWALL.

Skeelduck, also spelt **Skielduck.** A Scottish name for the Shelduck; cf. SKELDUCK, SKEELING GOOSE.

Skeeling Goose. An obsolete Orkney name for the Shelduck from Sibbald 1684, with which compare the lost Scottish **Skilling** given as a Goose name in 1578; see under ROUTHEROCK, also 1553 'skilling guse'. The qualification GOOSE is not unacceptable for the largest of our native Ducks, while *skeeling* consists of the suffix -ING redundantly added to *skeel-*, otherwise seen in synonymous Scottish SKEELDUCK, also spelt SKIELDUCK, in respect of the root vowel contrasting with Northern English SKELDUCK, closer to the standard form SHELDUCK.

Skeer Devil, see DEVIL 2.

Skelduck. A Cumbrian and Northumbrian name for the Shelduck, showing dialectal *skel-* for *shel-*; cf. SKEELDUCK.

Skiddaw. An East Lothian name for the Guillemot, echoing the shrill voice of the juvenile bird. Cf. KIDDAW.

Skiddy, also **Skiddy Cock.** West country names for the Water Rail, based on *skid* (a thinned variant of *scud*) run quickly, known since 1787 'Skiddey or Skiddey Cock'. A variant of SKITTY.

Skielduck, see SKEELDUCK.

Skiff. A Sussex name for a Tern, echoic in origin and as such comparable to the synonymous KIPP; see further under TERN.

Skilling, see SKEELING GOOSE.

Skirl Cock. A Derbys. name for the Mistle Thrush, alluding to the harsh cry, and close to synonymous SHRILL COCK.

Skirr. A name for a Tern from Lamby Island (north of Dublin Bay), imitative of the screaming call, and most akin to KIRR MEW; see further under TERN.

Skitty, also **Skitty Cock.** Terms for the Water Rail from Dorset and Somerset, Devon and Cornwall, in Somerset also meaning Moorhen, in Devon also Spotted Crake. The etymology of the name shows that the last two meanings are derivative, for Skitty is based on *skit* run quickly, cf. Devon 'Her rinn'th like a skitty' (*EDD*). The names refer to the Water Rail's habit of darting from one piece of cover to the next; cf. the variant SKIDDY, further synonymous RUNNER. Such behaviour is not characteristic of the Moorhen or the Spotted Crake.

Skooie, see SKUA.

Skua. This exotic word first appears in print in the Latin of Clusius 1605 denoting the Great Skua. The information came from Højer of Bergen, Norway, who *c.*1603–4 had got to know the species from the Faroe Islands together with its local name *skúvur* nominative, *skúv* accusative. This he reproduced as *sku*, whence the Latinized *skua*. Ray 1676 (Latin text) follows Clusius, taking over the Latinate form unchanged in the English version of 1678 'the *Skua* of *Hoier*'. Pennant 1768 noted the name, in 1776 using 'Skua, Skua Gull'. This bird Fleming 1828 renamed 'Common Skua', so converting Skua from a spe-

cific name into a generic term, and this usage was followed by subsequent writers.

The ancestor of the Faroese word is Old Norse *skúfr*, which also means tassel. The name is essentially imitative of the pursuit cry, but interpreted as though the bird was calling tassel. The Norse term also survived until recently in Shetland as **Skooie,** where -IE is a secondary diminutive. Since it is unlikely that the Great Skua was known to the Vikings before they colonized Orkney and Shetland towards 800 AD, the Norse name would actually arise in this area at that time. See BONXIE.

Skuttock. A Scottish name for the Guillemot, made out of onomatopoeic *skutt-*, reproducing the call of the immature bird (cf. SCOOT), and the suffix -OCK.

Skylark. First seen in Ray 1678 'The common or Skie-Lark'. One is surprised to discover that this most English-sounding of names is not really English at all, but a translation of Gesner 1555 *Himmellerch* (his own Swiss German). The genuine, traditional English name is LARK pure and simple. Pennant 1768 'Skylark' recognized the usefulness of the new word as a specific term, thus effectively giving it standard status.

Slavonian Grebe. A name first appearing in Montagu 1802 'Horned or Sclavonian Grebe' with the remark 'Dr. Latham says that it is found in Sclavonia', a reference to the north Russian home of the European birds. Jenyns 1835 and then Yarrell 1843 adopted the name, which now became the more usual term, found e.g. in BOU 1883. Later it tended to be given in the modern spelling, as in Hartert 1912.

Small Straw, also **Straw Small.** Yorks. names for a Whitethroat or other Warblers, with a corrupt form **Smastray** also found in Ches. Such birds build their nests with fine pieces of hay, small bits of straw, etc., hence the present names. For a similar formation, see WINDLE STRAW. Cf. JACK STRAW, also HAY JACK.

Smatch. A term for the Wheatear, first in Turner 1544 'Caeruleo a clotbird, a smatche, an arlyng, a steincheck', whence Merrett 1667 'Caeruleo Clot Bird, Smatch, or Arling, a Stone-check'. The name is imitative, cf. synonymous SMITCH; it is apparently unknown today except in the compound HORSE SMATCH.

Smew. Known since Charleton 1668 'Boscas Mergus, the Diving Widgeon, in Norfolcia the Smew'. It occurs in Ray 1678 (index), but he preferred WHITE NUN. Pennant 1768, however, chose the present term, de facto establishing it as the standard. This Norfolk word, also recorded as **Smee** or **Smee Duck**, imitates the whistling sound heard from this species. As seen above, it can also denote the WIGEON, another well-known whistler; see further under this name.

In spite of the late attestation, the name appears to be traditional, cf. the Dutch name for this bird smient (-ent duck), also local German Schmeiente.

Smitch. A Sussex term for the Wheatear, onomatopoeic, cf. synonymous SMATCH. See FALLOW SMITCH.

Snabbie. The Chaffinch is so called in Dumfries and Kirkcudbright. Origin unknown.

Snawful, see SNOW FOWL.

Snent, see STINT.

Snipe. Goes back to Middle English snípe, first noticed c.1325; the form occurring in Old English is, however, sníte. The explanation of the p-type in later English has been regarded as uncertain. Firstly, it is not unthinkable that a (fortuitously unrecorded) *snípe actually did exist in Old English times. Secondly, the type could have developed under the influence of synonymous Old Norse mȳrisnípa; see MIRE SNIPE. But there are no doubts that the p-type is the original one, as the Continental cognates prove; see SNITE. It presupposes a Proto-Germanic stem

*snípōn- with the meaning long, thin object. The term thus basically denotes the Snipe's characteristic bill; in other words, it is a part for whole name, as e.g. SHAG. This being so, it is even easier to understand why the name was applied to the Woodcock as well as the Snipes proper, as indeed, in popular usage, it still often is. Otherwise Snipe is to be taken as denoting the Common Snipe. Both Merrett 1667 and Ray 1678 have 'Snipe or Snite', but Pennant 1768 de facto decided the issue by choosing the former; see COMMON SNIPE.

Local names include HORSE GOWK and HEATHER BLEAT.

Snite. This now only very local term for the Snipe, confined to the extreme south-west, is in fact the oldest recorded name for this bird, Middle and Old English sníte, first occurrence c.725. The present word, unparalleled in the other Germanic languages, must surely be secondary, since only forms with p, never t, are to be found there, as Dutch snip, Middle Dutch snippe, sneppe, German Schnepfe (regularly from Old High German snepfa, where pf has evolved from original pp), further Old Norse -snípa, see MIRE SNIPE. Similarly Old English wudusníte woodcock corresponds to Dutch woudsnip or German Waldschnepfe with the same meaning. See SNIPE.

The origin of the present name has been an unsolved puzzle. We refer, however, to wudusníte woodcock above, and further note an early attestation of the same sense for the simplex: c.1000 'snite uel wodecocc'. As explained under WOODCOCK, there is good reason to believe that superstitious avoidance of the proper name of a much sought-after species played a part in the creation of names for this bird. It is therefore not unlikely that Old English sníte is the result of the deliberate deformation, in the service of name taboo, of original snípe, which by pure chance has not come down to us in the surviving monuments of Old English. For an example of comparable name deformation, see MOTHER CAREY'S CHICKEN.

Snow Bird. A name for any bird appearing in snowy weather, known since 1688. It is, in practice, most commonly used in the north of England and in Scotland as a local name for the Snow Bunting. Cf. SNOW FOWL.

Snow Bunting. Coined in 1771 in reference to the bird in North America (*OED*), rendering Linnaeus 1758 'Emberiza nivalis'. The name was taken up by British writers, and although systematics eventually changed, the vernacular coining has remained, cf. BOU 1883 'Snow Bunting *Plectrophanes nivalis*'. For genuine folk-names, see SNOW BIRD, SNOW FLAKE.

Snow Flake. Widely known, as a Scottish term especially, for the Snow Bunting, first seen in Pennant 1770; it is a corruption of SNOW FLECK.

Snow Fleck. A term for the Snow Bunting, first noticed in 1683. It is the antecedent of the now much better known SNOW FLAKE which has largely replaced it. Nevertheless, the present form survives in north-west Scotland, and sporadically in England, being recorded from Norf., Sussex, and Hants. The bird takes the name from its appearing in snowy weather and from the brown flecks or markings on the white plumage.

Snow Fowl, commonly in Scottish spelling **Snawful.** A name for the Snow Bunting from Shetland and Caithness. A word uniquely found in this area will be a relic of Norn, the ultimate source being Old Norse *snjófugl* (attested in the variant *snæfugl*) literally snow bird.

Snowy Owl. The name of this winter visitor to the far north of Britain is first found in Latham 1781 who commented 'the whole plumage is as white as snow'. The term won immediate acceptance.

Snyth. An Orkney name for the Coot, a

Norn word denoting the white frontal shield, cf. Norn *sny* white spot on a horse's face, doubtless the same word with loss of final *th*. Antecedents uncertain. There is similar motivation in BALD COOT.

Solan, usually **Solan Goose,** formerly **Soland, Soland Goose.** An originally Scottish name for the Gannet, properly Solan, first recorded *c*.1450 in the corrupt form Soland. A name of this shape could scarcely fail to lead to the tautological Soland Goose (first 1536 'soland geis'). This form was adopted by English naturalists, as Turner 1544 'Solend Goose', Merrett 1667 'Soland Goose', likewise Ray 1678 and Pennant 1768, and remained in common use (despite Pennant 1776 'Gannet') until the middle of the 19th cent., thus Yarrell 1843 'The Gannet or Soland Goose', though he regarded the former as the main name. The alternative form Solan Goose is of comparable age (first 1583 'solane geis'); it was used by Charleton 1668 and Martin 1698, and during the latter half of the last century it more or less replaced Soland Goose, but admittedly as second fiddle to GANNET.

Although the simplex Solan is not recorded as such until 1749, it is the primary form of the name. It represents the Lowland Scottish pronunciation of a lost Gaelic *sùlan, the term once used for the Gannets on Ailsa Craig in the Firth of Clyde. The Gaelic word, in its turn, was borrowed from Old Norse *súla*, whence in the dialect of St. Kilda *sùl*; our form shows the secondary ending -*an*, productive in Gaelic bird names, cf. PTARMIGAN, TRILLACHAN.

The literal meaning of the name is cleft stick, in reference to the crossed wing-tips, black in contrast to the pure white of the rest of the plumage, and so conspicuous when the birds are standing or sitting. Such a name will then be due to fowlers visiting the gannetries, a term vague enough to function as a noa word. As far as is known, the Norsemen could not have become closely acquainted with the bird until they reached the gannetries in the north and west of Scotland at the beginning of the Viking Age, towards 800 AD, at which time the name would be coined. Old Norse *sūla* literally cleft stick is itself a corruption of an original form *svala*, which had also, at a very much earlier date, become the name of a quite different bird; see SWALLOW. *ZAA* xxi 118–20 (1973), *Fróð.* xxiii 26f. (1975), *SGS* xii 275f. (1976)

Song Thrush. An entirely appropriate name for one of the finest of our songsters, first recorded by Merrett 1667. The name flourished, indeed perhaps actually arose, in those districts where THROSTLE was not in use; see THRUSH. It may be said to have been effectively the standard name for this species since White 1770; see THROSTLE.

Spadger. A playful alteration of SPARRER, i.e. (HOUSE) SPARROW, heard almost everywhere in England; in Oxon. and Worcs. by suffix change also **Spadgick**, see -ICK. For another jocular form, see SPUR.

Sparr. Recorded from Kent and Sussex as a term for the House Sparrow. It is to all appearances a survival of an ancient collateral form of SPARROW without the *w* element, cf. rare Middle English *spare*, Old English **spara*, *speara*, corresponding to Old High German *sparo*, Middle High German *spare*, Modern German *Sperling* (with diminutive suffix producing vowel change).

Sparrow. A traditional name for various passerines and similar-looking birds, this general sense also in SPARROW-HAWK. Otherwise the word figures in the standard names of three British birds: HEDGE, HOUSE, and TREE SPARROW, while locally it may be used of other species, e.g. REED SPARROW Reed Bunting. It goes back via Middle English *sparewe* to Old English **sparwa*, *spearwa*, of Germanic ancestry, cf. Middle High German *sparwe*, further Old Norse *spörr* (see SPUR), Gothic *sparwa*; see also SPARR.

To represent a dialect or sub-standard pronunciation of Sparrow, the spelling **Sparrer** is occasionally used; cf. SPAD-GER.

Sparrowhawk. Going back via Middle English *sperhauk* to Old English *spearha-foc*, the species is appropriately named since it preys largely on Sparrows i.e. small birds; see SPARROW. To all appearances the name arose in this country, for it is not found in the other West Germanic languages, although these do use a term based on the word Sparrow. It is met with, in Latin dress, as early as the Lex Salica (end of 5th cent.), as *speruarius*, cf. Old High German *sparwāri* (becoming Modern German *Sperber*). Old Norse has, it is true, *sparrhaukr*, but this will hardly be an original Norse name, but one adopted from Old English by Viking falconers learning their art in England; see also under GOSHAWK. Cf. BLUE HAWK, HEDGE HAWK.

Speck. An East Anglian word for Woodpecker, known since 16th cent. 'Specke'. It is a borrowing from Norman French **spec*, *espec*, itself a loan from medieval German *spech*. This latter preserves the more primitive form of the Common Germanic name, otherwise found with secondary final *t*; see SPEIGHT. One postulates Proto-Germanic **spihaz*, plausibly akin to synonymous Latin *pīcus* (initial *s* before consonant being optional, the so-called *s* mobile). Also WOOD SPECK.

Speight. A word for Woodpecker, recorded between 1450 'Specht' and 1656 'Speight'. The term was early used as a personal name, the record going back via 1297 'John Specht' (Reaney 1976) to Old English *Speht* attested in the Dorset village name (Domesday Book, 1086) 'Spehtesberie', today *Spettisbury* literally Speight's Bury. It is the Common Germanic name, corresponding to Dutch *specht*, German *Specht*, further Old Norse *spætr*; the consonant *t* is, however, not original, see SPECK. The term survives in WOOD SPEIGHT and corruptly in SPRITE.

Even though traditional, the present term was evidently never prominent here, as it is on the Continent. In contexts where the exact meaning can be ascertained, the name is seen to refer to the Green Woodpecker, as indeed is usual with any popular Woodpecker name.

Spensie. A Shetland (Yell) name for the Stormy Petrel, origin unknown, but apparently formed with the Norn suffix -si, cf. BONXIE.

Spink. A local name for the Chaffinch, recorded from many parts of the country, both north and south, attestation going back to c.1425. The word is unquestionably of onomatopoeic origin, like its synonym PINK It is not found in the other West Germanic languages, but is quite widespread in Continental Scandinavia, though unattested in Old Norse, perhaps fortuitously. Spink may therefore not be a spontaneous English creation, but have been introduced by the Danes in Viking times. It is even possible that we then have a very ancient name. A Proto-Germanic root *spink- presupposes Indo-European *sping-, met with in the Greek Chaffinch name spíngos, and parallel to *ping- which lies behind FINCH.

Spink may replace Finch, see e.g. BULL SPINK.

Spite. A way of spelling SPEIGHT.

Spoonbill. After Charleton 1668 had called this bird 'spoon-bill'd heron', Ray 1678 introduced the present term 'The Spoon-Bill. *Platea sive Pelecanus* . . . The Bill . . . is of the likeness of a Spoon whence also the Bird itself is called by the . . . Dutch *Lepelaer*, that is, Spoonbill'. Pennant 1768 took up the name and his successors followed suit. For traditional names, see POPELER, SHOVELER.

Spotted Crake. It seems unlikely that this rare and furtive species ever acquired a folk-name of its own. Pennant 1768 called it 'Small Spotted Water Hen', alluding to the white spots scattered over the plumage, but in later editions altered this to 'Spotted Gallinule'. Bewick 1797 called the bird the 'Water Crake'. Finally, Shaw 1824 in an apparent desire for the best of two worlds, created yet another name, Spotted Crake, which subsequent writers mercifully accepted.

Spotted Flycatcher. A term coined by Pennant 1776 and taken up by his successors, the spots being the striations on the head and underparts. Traditional names include BEAM BIRD, CHAIT, CHERRY CHOPPER, POST BIRD, WALL BIRD.

Spotted Woodpecker. An abbreviation of GREAT SPOTTED WOODPECKER, not infrequently found during the past fifty years or so. Such usage becomes possible and tolerable when BARRED WOODPECKER is used instead of LESSER SPOTTED WOODPECKER.

Spowe. An obsolete Norf. term for the Whimbrel, first recorded in 1519. It derives from Old Norse spói. In the Scandinavian languages, however, the word is also used for the Curlew. This will be the original sense, for a bird name of this shape is without doubt imitative, as such fitting the call of the Curlew rather than the tittering cry of the Whimbrel; see further under these names. The extended use of the term to include the Whimbrel is, of course, explained by the general similarity of the two closely related species. Cf. HALF CURLEW. One naturally assumes that both meanings were once current in England, too.

Sprite. A Suff. term for the Green Woodpecker, properly SPITE=SPEIGHT.

Spur. A Scottish term for the House Sparrow, apparently a borrowing from Old Norse spörr, ultimately the same word as SPARROW. There is a derivative form **Spurg**, explained as a contraction of an (unrecorded) diminutive *Spurock* (see -OCK), by metathesis becoming **Sprug**, whence the jocular **Spug**, commonly **Spuggy**, well known in England too, above all in the north.

Spurre. An Ulster name for a Tern, onomatopoeic; cf. PURRE 2, and see further under TERN.

Squacco Heron. In the absence of a native name for this rare vagrant, Willughby noted *c*.1672 'Ardea quam Sguacco vocant in Vallibus dictis Malalbergi'. This he had from Aldrovandi 1603, Ray 1678 translating 'the Heron which they call Sguacco in the Valleys of Malalbergo'. Hill 1752 spells the name 'squacco', whence Latham 1785 (adding the generic *Heron*) 'Squacco Heron' which has since prevailed. As Ray supposed, this—purely local—Italian name conceivably imitates the husky voice.

Squeaker, Squealer, see SCREECH MARTIN.

Stag. A Norf. name for the Wren. It goes back to the Vikings, cf. Old Danish *stag* stick, here denoting the cocked-up tail; see further under WREN.

Stanchel. A north country and Scottish name for the Kestrel, first *c*.1450 'stanchal'. The name doubtless descends, albeit irregularly, from Old English *stängella* and is thus a variant of STANIEL, etc.

Stand Hawk. A Yorks. (West Riding) name for the Kestrel. This clumsy-sounding name, clearly meant to allude to the Kestrel's hanging stationary in the air, looks suspiciously like an explanatory corruption of STANNEL HAWK.

Stane Putter. An Orkney name for the Turnstone, not Norn, but Scottish in form and therefore a relatively recent name; it means stone pusher.

Staniel, see STANNIEL.

Stank Hen. A local Scottish name for the Moorhen, known since 1766, *stank* meaning pond.

Stannel. A northern name for the Kestrel, first in 1601 'Kestrill, or Stannell', arising from older 'stanyel'; see STANNIEL. It may become **Stannel Hawk**; see STAND HAWK.

Stanniel. Recorded in 1838 as a provincial name for the Kestrel. It is a conflation of STANNEL and Staniel, a 17th-cent. form seen again in Late Middle English *c*.1475 'stanyel', which in turn has a formal antecedent in Old English *stängella* literally stone yeller, where *stone* may well be an intensifying epithet (as in stone deaf). In Old English, however, the word occurs only as a translation of Latin *pelicanus*. This may well have been a reasonable approximation, because in German a similar word *Steingall*, also diminutive *Steingellel*, is known to have been used for the Green Sandpiper. But in view of the basic meaning of the name, there is no reason why it should not have denoted the shrill-voiced Kestrel as well, one sense prevailing in certain districts, the other elsewhere, as not infrequently happens with bird names; see e.g. BARLEY BIRD. There is, at any rate, little doubt that here is the oldest English name for the Kestrel.

Stare. The primary form of STARLING, recorded in local use from most parts of the country. It goes back to Old English *stær*, with Continental cognates in Old High German *staro*, Modern German *Star*, Old Norse *stari*, indicating Proto-Germanic *staran*, this in turn related to Latin *sturnus*. The name must thus have been in existence as early as the second millennium BC. It has been formed from an imitative root *star-*.

The present word competed with its diminutive Starling, thus Merrett 1667 'Stare . . . Sterlyng', Ray 1678 'Stare or Starling'. Pennant 1768 preferred 'Stare' and Latham 1787 shared his preference, but in general this form was not used by their successors. Selby 1833 was an exception, after which the name becomes rare in writing, where it is now archaic.

Starling. Goes back to Old English 11th cent. *stærling* young starling, formed with the suffix -LING from the primary name *stær*; see STARE. The original diminutive sense was later lost and the present form is seen to compete with Stare as the ordinary name for the species, in part replacing it in local use, and in the standard language eventually superseding it altogether. It is the only form found in Turner 1544, similarly in Albin 1738 and Edwards 1743 and most of the early 19th-cent. naturalists, Yarrell 1843 finally confirming its standard status.

In appearance, the immature Starling differs appreciably from the adult, and this led not only to the creation of the present name, but also of certain distinctive names for the juvenile, as BROWN or GREY STARLING; conversely the mature bird may be distinguished by a suitable epithet, e.g. BLACK STARLING.

The name Starling is often locally pronounced, and occasionally written **Starlin**; from this form developed the variants STARNIL and STARNEL.

Starn. A Norf. expression for the Black Tern. It occurs, in Latin garb, in Turner 1544 'sterna', which Linnaeus 1758 adopted as the generic name for the Terns. Fortuitously absent from the records of Middle English, the word reappears in Old English *stearn*. It is echoic in origin, closely akin to TERN. Turner describes the species as abundant and noisy; it remained a familiar bird in the Fens until the middle of the 19th century.

Starnel. A name for the Starling found at various places in the north country and southwards as far as Lincs. and Worcs. It is occasionally recorded in a more archaic form **Starnil**, which will have arisen by metathesis from **Starlin**, a common pronunciation of STARLING; see -LING.

(-start. Formerly an independent word meaning tail, seen in Middle English *stert*, Old English *steort*; it is of Germanic age, cf. Dutch *staart*, Low German *Steert*, locally (High) German *Sterz*, Old Norse *stertr*. The independent word disappears from English sources during the 13th cent., being replaced by *tayl*, but survives as a component of certain bird names, cf. REDSTART, WAGSTART, WHITESTART. As a term for a bird's tail, start is much older than TAIL.)

Steer. A variant of STARE occurring from Shrops. to Glos.

Steingall. Known from Turner 1544 'kistrel, kestrel, steingall'. The word has been seen as an (admittedly obscure) variant of STANIEL. We regard it as non-English, for the following reason. After giving his English names, Turner makes a practice of adding the German equivalents. But, in the present instance, no German is quoted, or so it appears at first sight—until one realizes that 'steingall' is properly the German name corresponding to 'Kistrel, Kestrel' which, through a printing error, is made to look like an English name. Indeed, there is such a German name, albeit otherwise recorded as denoting a different bird; see under STANNIEL. See also STONEGALL.

Stint. A local term for the Dunlin, in use along the east coast from Northumb. to Suff., also in Sussex. It makes its debut in 1452 'Styntis', the present spelling first in 1622. We propose to regard the word as a folk-etymological alteration of **Stent**, known from a document of 1579, and this in turn as a corruption of **Snent**, the name used in Berwicks. The ultimate meaning we assume to be beak, arguing as follows.

The Dunlin with its long, strong bill rather closely resembles the Snipes, hence such local Dunlin names as **Sea Snipe**, common in the north and in Scotland, or **Little Snipe**, heard in Yorks. (Swaledale), further Glos. **Sand Snipe**. We recall that SNIPE literally means beak, and suggest that Snent belongs to the same series of words, to which one may, indeed, add NEB in view of its Low German congener *sneb* (on initial *s* see under SPECK).

The present name has passed into the standard nomenclature in the term LITTLE STINT.

Stock Dove. Known since *c*.1340 'Stokdowe'. *Stock* here means tree-trunk, alluding to the nesting place, a hole in a tree-trunk, which distinguishes this bird from the allied species; also called HOLE DOVE. See further under ROCK DOVE, WOOD PIGEON.

Stock Duck. An Orkney and Shetland name for the Mallard. It is a part Anglicization of a lost Norn expression corresponding to the common Scandinavian term for this species, e.g. Icelandic *stokkönd* (-*önd* duck, cf. ENDE); the term also occurs in German *Stockente*. The precise meaning of *stock* in this connection is unknown.

Stone Chack, Stone Chacker. Stony wastes are the Wheatear's natural surroundings, where a behavioural pattern has led to name-giving, namely the habit of flitting from stone to stone, there to perch awhile, bow the head, and flirt the tail. Accordingly, a word meaning stone is of frequent occurrence in Wheatear names in any language. At the same time, certain calls play a part in the nomenclature. One of these is represented by **Chack** (variants **Check**, CHUCK) or CHAT; another call is reminiscent of lip-smacking, hence SMATCH, SMITCH. These calls are, however, not unlike sounds produced by our two species of *Saxicola*, and the selfsame imitative syllables may enter into names for these birds, too. As a result, Wheatear names based on these calls are usually qualified by *stone* in order to distinguish them from *Saxicola* names, hence chiefly north country and Scotland Stone Chack, Stone Chacker, also **Stone Check** (first in Turner 1544 misprinted 'steinchek', Merrett 1667 correcting 'Stone-check'), **Stone Checker**, Sussex **Stone Chuck**, **Stone Chucker**,

then north country STONECHAT, STONE CHATTER; see -ER.

Although the above-mentioned names properly denote the Wheatear as opposed to *Saxicola*, confusion in naming these species is nevertheless widespread and transferences of sense are often found, the most notable example being STONECHAT; see further under this name.

Stonechat. After Pennant 1768 'Stone Chatterer', apparently based on the living folk name **Stone Chatter** which goes back to Ray 1678 'Stone-smich, or Stone-chatter', we find Tunstall 1771 using the short form 'Stone Chatt' presumably to match 'Whin Chatt'. This was taken up by Latham 1783 and established itself as the standard. The name originally denoted the Wheatear, a meaning preserved locally in the North, as explained under STONE CHACK.

In general, popular names for the Stonechat are equally valid for the WHINCHAT; see under this name.

Stone Curlew. First recorded by Merrett 1667 'Stone Curliew', this rather scarce summer visitor acquired its name in Norfolk, as mentioned by Ray 1678 under 'Stone-Curlew'. The motivation is evident: the bird's wailing 'coo-lee', heard at night when the bird is particularly active, vividly recalls the voice of the commoner, resident Curlew, while the distinguishing epithet *stone* was suggested by the habitat—barren stony tracts on heaths and downs. This traditional name had, however, to compete with terms coined by naturalists (see GREAT PLOVER and NORFOLK PLOVER) not emerging as the standard until BOU 1883.

Stonegall. A name due to Merrett 1667 'Stannel or Stonegall, Kestrel or Kastrel'; it was quoted by Ray 1678 'Kestrel, Stannel, or Stonegall'. A purely book word, simply an innocent normalization of Turner's (German) STEINGALL.

Stone Hatch. A Suff. name for the

Ringed Plover. The slight depression in the sand, which serves this bird as a nesting site, is often lined with pea-sized stones; upon these the eggs are laid and hatched out.

Stork. An ancient name going back to Old English *storc*, cognate with German *Storch*, further Old Norse *storkr*, the forms presupposing Proto-Germanic **storkaz*. The term was originally a nickname, for it simply means stick—there are reminiscences of this primary sense in German dialect. Where the Stork nests on roof-tops, as it commonly does, it may often be seen in a characteristic pose—standing on one leg, i.e. on its 'stick'.

This etymology has a sequel. In older German dialect, *Storch* could have the secondary meaning of penis, this evidently being the source of the (hitherto unexplained) fable that the Stork brings the babies.

Fród. xxii 111f. (1974), xxiv 76f. (1976)

Storm Cock. Unlike most other birds, the Mistle Thrush ignores the trials of wind and wintry weather and, conspicuous on some leafless branch, pours forth its song heedless of the storm. The term is widespread, but in Scotland is commonly applied to the Fieldfare, most typically a winter bird.

Storm Petrel, Stormy Petrel, see PETREL.

Straw Mouse. A Ches. term for a Whitethroat, with correspondences in Yorks. **Straw Small**, Westmor. **Straw Smear**, and Rutland **Straw Sucker**. The epithet *straw* refers to the common nesting material; for *mouse* see TIT-MOUSE, and for *small* see SMALL STRAW. The word *smear* is surely a corruption, but not impossibly associated with *sucker*, which derives from an ancient name type discussed under HAY SUCK.

Strannie. An Ayrshire term for the Razorbill, first noticed in Montagu 1802 'strany' as a name for the (Common) Guillemot. It represents (unrecorded)

Scottish Gaelic **strannag*, as confirmed by Manx Gaelic *stronnag* guillemot. The name was motivated by the call; it is based on Scottish *strann*, Manx *stron* snort, snuffle. In the Anglicized form, the diminutive suffix -IE has been substituted for the corresponding Gaelic -ag. The sense Guillemot is primary, for the reasons given under MURRE.

Stumpit. A Lancs. name for the Wren, a dialect variant of **Stumpy** also used as a Wren name. Cf. CRACKY.

Stumpy Toddy. A Ches. name for the Wren, *stumpy* naturally meaning short and squat, *toddy* meaning tiny. Cf. TITTY TODGER.

Summer Snipe. Described by Newton 1896 as the most popular English name for the Common Sandpiper. It is used in the Fens and in various places to the north, becoming general in Scotland.

Summer Teal. A Norf. name for the Garganey, known since Charleton 1668. This bird, so similar to the resident Teal, is of course a summer migrant to East Anglia.

Surf Duck. A Scoter name used in Scotland. These Ducks feed just off shore, hence the epithet *surf*.

Surf Scoter. A name coined by Fleming 1828 for this rare vagrant, using SCOTER generically and taking specific *surf* from the popular Scottish Scoter name SURF DUCK—Fleming was a Scotsman.

Swaabie, etymological spelling **Swarbie**. A Norn term for the Greater Black-backed Gull from Orkney and Shetland. It is a reduced form (see -IE) of **Swart Bak**, occasionally found in formal writing and close to the ancestral Old Norse *svartbakr* literally black-back. Cf. BAAKIE.

Swallow. A traditional name, Middle English *swalowe*, Old English *swealwe*, of Germanic age, cf. Dutch *zwaluw*, German *Schwalbe*, also Old Norse *svala*. The forms presuppose Proto-Germanic *swalwō, the literal meaning of which was cleft stick, the name having been motivated by the characteristic forked tail. See SOLAN.

Not surprisingly, the name of the Swallow was often extended to include the similar and closely related Martins, and also the Swift. As a result, Swallow is a constituent of many local names for these species. See also CHIMNEY SWALLOW.

Fróð. xxx 105f. (1982)

Swan. A name unchanged in spelling since Old English times. It is of Germanic age, cognate with Dutch *zwaan* and German *Schwan*, further Old Norse *svanr*, all presupposing Proto-Germanic *swanaz*. The word is etymologically identical with Sanskrit *svanás* and Latin *sōnus* from *swonos*, both meaning sound, noise. The present name thus belongs to an early stratum in the nomenclature, when an abstract noun could be transformed into a bird name, cf. LARK, LOON.

The Proto-Germanic language already had a generic word for Swan, inherited from Indo-Germanic times, see ELK. Conceivably, then, the present name arose to denote a particular species. We may ignore the inconspicuous Bewick's Swan and concentrate on the possible claims of the remaining species, the Whooper and the Mute. At first, the Whooper seems the obvious candidate. But then one may wonder how a word having the general sense of sound could refer to the voice, since virtually all birds have voices and hence make sounds. The word therefore plausibly refers to some other, distinctive aspect calculated to lead to name-giving. We now recall that the flight of the Mute Swan is remarkable for its singing, metallic throb which, in favourable conditions, is audible a good mile off and more. Such a unique phenomenon is more than likely to inspire a name. We therefore identify the name Swan in the first place with the Mute Swan.

Swarbie, Swart Bak, see SWAABIE.

Swift. Originally just one of a host of local names for this species, it first comes to notice in Charleton 1668 'Horse-Marten, or Swift', then in Ray 1678 'The black Martin or Swift'. Pennant 1768 took the present name and may be said to have standardized it, as later naturalists generally followed his usage. In spite of its relatively late appearance in the records, this name could be very much older; indeed, the now peculiar employment of an adjective as a singular noun bespeaks an origin in medieval English, when any adjective could automatically function as a noun, cf. DUN, REE (under REEVE). The self-evident motivation is seen again in the local synonym SWING DEVIL.

The Swift has been regarded as a bird of ill omen, a superstition which gave rise to many names, see DEVIL 2.

In view of its superficial resemblance to the Swallow and the Martins, the Swift is often called after them, hence SWALLOW and MARTIN are of not uncommon occurrence in Swift names.

Swine Pipe. A name for the Redwing, often quoted in bird books and first occurring in 1668. The Redwing may take its name from the impressive whining note, hence it may be called WHIN THRUSH from Middle English *whīnthrushe*. The note in question, here reproduced as *whīn*, could equally well be represented by *swīn*, a close parallel being seen in such imitative pairs of synonyms as KIDDAW, SKIDDAW, or TERN, STARN. As the ancestor of the present name, we therefore posit Middle English *swīn pīp*, the second word also onomatopoeic, whence in the course of the regular phonetic evolution of the language the name as it is known to us today.

Swing Devil, see DEVIL 2.

Swing Tree. Given as a Suff. name for the Goldcrest, perhaps originally *Swing-in-tree*, alluding to the nest which is suspended by 'handles' from a branch.

Sycock. A Derbys. and Notts. name for the Mistle Thrush. An alternative spelling would be *Saicock, which suggests original *Saith Cock, cf. synonymous SAITH.

T

Taggy Finch. A Worcs. term for the Chaffinch, the significance of *taggy* being unknown.

(Tail. Continues Middle English *tayl*, Old English *tægel*, cognate with Old High German *zagel* tail, further Old Norse *tagl* horse's tail, the latter a more original sense, cf. Gothic *tagl* hair. The surviving records of Old English show the term in use for the tail of any animal, but we have not noticed an instance referring to a bird's tail, and there is evidence which suggests that the word was, in fact, not yet so employed; see -START. In the Middle English period, however, the present word came into use in that sense. The first precisely datable occurrence of *tail* as a constituent of a bird name is not earlier than 1510 (see WAGTAIL), but such formations can doubtless go back to Middle English times; see further under WHITE TAIL.)

Tame Swan, see WILD SWAN.

Tammie Norie, see NORIE.

Tangle Picker. A Norf. name for the Turnstone, *tangle* denoting a sort of seaweed, *picker* meaning pecker, grubber, this usage found again in PICKEREL. The word *pick* is a collateral form of *peck*, locally having the same meaning, and general in the expression *pick at one's food*.

Tang Whaup. Literally seaweed curlew, an Orkney and Shetland term for the Whimbrel.

Tapper. Given as a Leics. name for the Lesser Spotted Woodpecker, but without a doubt generic rather than specific.

Tarrock 1. A northern, chiefly Scottish name for the (Common or Arctic) Tern, known since 1795. It consists of *tarr-*, a syllable echoing the harsh, shrill cry, and the well-known suffix -OCK. Names for these birds are commonly onomatopoeic; see TERN for other examples.

Tarrock 2. A west Cornwall name for the immature Kittiwake, first recorded by Ray 1674. It was regarded by the early naturalists as denoting a separate species and thus obtained a certain prominence in ornithological writing. A bird name from this locality and of this age must surely be a relic of the former Cornish language. Seeing that Kittiwakes are such clamorous subjects and their various names most commonly imitative in origin, it is very likely that the present name is also onomatopoeic. But it is, of course, independent of TARROCK 1, with which it is only fortuitously identical.

Tartan Back. A name for the Brambling, locality not given, motivated by the striking, brindled plumage.

Tatler. A Sandpiper name, locality unspecified. The word, literally babbler (cf. *tittle-tattle*), naturally alludes to the voice, as so often in Sandpiper names.

Tawny Owl. Coined by Pennant 1768 as more appropriate than the already existing, genuine folk-names BROWN OWL or GREY OWL, and used by him to contrast with WHITE OWL, his term for the Barn Owl. The present name won the general acceptance of other ornithologists.

Teal. The first record dates from 1314 'teles', later attestations include Palsgrave 1530 'teele' and *c.*1532 'teyle', 1538 'teale', Turner 1544 'Tele', 1614 'Teils', the present spelling in Merrett 1667, again in Ray 1678 and Pennant 1768. There was in this case no competition for standard status from synonyms. Uniquely among common British birds, the Teal is not known to have ever been called by any other name than the present one. It may have entered, secondarily, into ATTEAL.

Though not met with in the surviving Old English sources, the name was certainly present either as *tēla (masculine) or *tēle (feminine); cf. Dutch and Low German teeling (with suffix -ing added). These names have been formed from a syllable tēl imitative of the chuckling sound heard from the birds when feeding. TPS 1975 183–6 (1977)

Teaser. A name for the Arctic Skua from the north-east coast, first in Pennant 1776 corruptly as FEASER. The name derives from the Skua practice of harassing other sea birds until they disgorge. Cf. FASKIDAR.

Teeting. A Pipit name from Orkney and Shetland, of Norn origin, consisting of onomatopoeic teet- and the derivative suffix -ING. Cf. TITLING.

Teeweep, Teeweet, Teewheet, see TEUCHIT.

Teistie, another spelling of TYSTIE.

Tell Pie. A north Yorks. term for the Magpie. The element tell refers, of course, to the chattering voice, cf. CHATTER PIE, here (we suppose) with the added nuance of tell-tale, talebearer.

Tern. First in Ray 1678 who describes it as used 'in the Northern parts'. There was doubtless a variant *Tarn arising as a result of common phonetic evolution (as explained under BERGANDER), cf. PICK TARNIE. As a formation, the name is closest to STARN, and onomatopoeic in origin like most names for this bird, as DARR, TARROCK 1, PURRE 2, SPURRE, KIRR MEW, SKIRR, also SCRAY, further KIPP, SKIFF, all inspired by the penetrating cry.

Ray 1678 regarded SEA SWALLOW as the most appropriate term, but Pennant 1768 took up the present name and his successors followed suit.

Teuchit. A Scottish name for the Lap-

wing, with literary status. It is first found in Holland c.1450 'Tuchet', next in 1549 'teuchitis'; a variant **Tewfit** occurs sporadically in the north of England. These forms derive from **Teewheet**, which in turn reflects onomatopoeic **Teeweet**, both surviving locally, the latter not unusual in the north of England, too. Another variant **Teeweep** occurs in Scotland. Like its better-known synonym PEWIT, the present name has every appearance of being traditional, cf. Low German Tiewit.

Thack Sparrow. Also given as **Thatch Sparrow**, a term for the House Sparrow known from Northants. and Shrops. Thack preserves the more original form, the primary sense being roof.

Thick Knee. A name for the Stone Curlew, first used by Leach 1816. It is an abbreviation of Pennant 1776 THICK-KNEED BUSTARD, not a folk-name, the thickening referred to hardly sufficing to attract the notice of the layman.

Thick-kneed Bustard. Pennant 1768 classified the Stone Curlew with the Plovers, calling it NORFOLK PLOVER, but in 1776 found it approached near to the Bustards, adding 'The knees thick, as if swelled like those of a gouty man'. In creating the present name, Pennant was following Belon 1555 who had constructed the term Oedicnemus out of Greek oidi- swollen and knēmē hollow of the knee, in allusion to a certain enlargement of the tibio-tarsal joint. See THICK KNEE.

Thirstle, see THRISSEL.

Thistle Finch. A name for the Goldfinch, occurring in Norf. and Worcs., also in Scotland, so called from its commonly feeding on thistle seeds. This very apposite name was first noted by Ray 1678 'Goldfinch, or Thistle-finch'. Pennant 1768 decided in favour of the former word, even though it was (and still is locally) also used for the Yellowhammer.

Thistle Tweaker. Literally Thistle

Plucker, this name for the Goldfinch (locality not stated) looks back, it seems, on a respectable ancestry. It apparently continues, albeit in a somewhat altered form, Old English *pisteltwige*. Here -*twige* likewise means plucker, today replaced by *tweaker* from a related, though not identical root.

Thistle Warp. A name for the Goldfinch known from sources dated 1606 and 1624. A word of this shape has every appearance of considerable antiquity, the second element being basically an agent noun associated with the verb *warp* to twist. This verb must, however, have developed a further sense of tweak, pluck, since the present name can hardly mean anything else than Thistle Plucker, precisely as THISTLE TWEAKER.

Thrasher. Today an American word denoting New World species, but originally a local British term for a Song Thrush. It is a variant of THRESHER, and will have been carried abroad by early settlers. It was never recorded in its native land, and is now forgotten here.

Thresher. A name for the Song Thrush. It consists of the suffix -ER attached to obsolete **Thresh** recorded *c*.1676, a name not occurring—perhaps fortuitously—in the Middle English records, but close to Old English *præsce*, itself akin to *prysce*, the linear antecedent of THRUSH.

The name is known from Berks., Bucks., and Oxon. From the same area comes **Thrusher**, evidently a modification of the present name under the influence of Thrush; see also THRASHER.

Thrice Cock. A Midlands word for the Mistle Thrush. A folk-etymological alteration of earlier **thraish cock* from Middle English **thrīshcoc*, cf. Middle English 'thrysche' (normalized *thrīsh*) going back to Old English *prȳsce*, see THRUSH.

A compound name comparable to the present is seen in Middle English *threstelcoc*, beside *threstel*, a variant of *thros-*

tel, see THROSTLE. As to the meaning, we recall that THRUSH often occurs locally in the specific sense of Mistle Thrush. See COCK.

Thrissel, etymological spelling **Thristle.** A secondary variant of THROSTLE, widespread in Scotland, also heard in Northumb. and Durham, further in Shrops., and again in Devon and Cornwall; in the last three counties there is also a form **Thirstle**, i.e. with metathesis (cf. BIRD). The present name is at least of Middle English age, being recorded as early as 1375 'thristill'.

Throg, Throggy. Pet names in Ches. for the Thrush, derived from THROSTLE; cf. the better known THROLLY.

Throlly. A widely used pet form of THROSTLE; see -IE.

Throstle. A traditional name for the Song Thrush, Middle English *throstel*, Old English *prostle*. It is closely allied to Middle High German *drostel*, further Modern German *Drossel*, Dutch *drossel*, initial *d* having regularly developed from original Germanic *p*. In these names, *l* indicates a secondary diminutive formation; it is absent from the related Old Norse *pröstr*. The Germanic name is thus akin to synonymous Latin *turdus*, root *turd-*, evidently for **trurd-* (i.e. with dissimilatory loss of the first *r*), from earlier **truzd* (such a change of original *z* to *r* being normal in Latin). Similar patterning is seen in the Slavonic cognate, e.g. Russian *drozd* from more original **trozd* (i.e. with assimilation of initial *t* to final *d*); and likewise in Lithuanian *strazdas*, with prefixed *s*, root *-trazd-*. Although the vowels in the root of this name are in part variable, the primitive consonant skeleton is uniformly *tr-zd-*, revealing an onomatopoeic formation, not unexpected in the case of this distinguished vocalist. In Germanic, these consonants have naturally been shifted to *pr-st-* (see the introduction, p. 7, and further under THRUSH).

The widespread distribution of the

consonant pattern, illustrated above, indicates that ancestral forms of this word will have been in existence as early as the second millennium BC.

In standard ornithological work, the present name has given way to its old competitor THRUSH, or more precisely SONG THRUSH. The earlier naturalists, however, usually quote it, thus Charleton 1668 'Thrush, Song-Thrush, or Throstle, or Mavis', Ray 1678 'Mavis, Throstle or Song-thrush'. Pennant 1768 actually chose it as his specific term, but he was not followed in this, and White 1770 speaks of the 'Song-thrush'. The name remains fairly common in general literature, but is today otherwise essentially a local, more northern word.

Thrush. A traditional name going back through Middle English *thrushe* to Old English *þrysce*, also with original long vowel *þrȳsce* (see THRICE COCK), together with the variant *þræsce* (see THRESHER) reminiscent of Old High German *drōsca*. It is of pre-Germanic age, witness Welsh *tresglen*, where *-en* is a singulative suffix common in this language, and *l* a familiar diminutive element, the remaining root *tresg-* implying a basic consonant pattern *tr-zg-*, regularly shifted to *þr-sk-* in the Germanic cognate. This consonant pattern shows Thrush to be simply a variant of THROSTLE with its Germanic consonants *þr-st-* from onomatopoeic, pre-Germanic *tr-zd-*, see under this name. We regard the *tr-zd-* pattern as the original, from which *tr-zg-* arose by dissimilation of *d* to *g*. To this there is a comparable, but independent, later development in the Slavonic languages: the typical form *drozd*, as in Russian, is occasionally changed to *drozg*, as in Slovene.

From the very beginning of our language, the present name has competed with Throstle, specifically Song Thrush. In districts where both names have been common, Thrush has generally been restricted in sense to Mistle Thrush, a state of affairs reflected in Turner 1544. In other parts, where

Throstle was not in use, the two species could be distinguished as SONG THRUSH and MISTLE THRUSH.

Given the degree of phonetic similarity between Throstle on the one hand, and Thrush and its ancient variant (above) on the other, it is not surprising that several hybrid forms have arisen, early examples being THRUSHEL and THRUSTLE.

Thrush Cock. A word for the Mistle Thrush occasionally found in the Midlands beside THRICE COCK.

Thrushel. A Shrops. word for the Song Thrush, first noticed as such in 1499 'Thrusshill or thrustyll', but earlier still in the compound 'Thrushylcok' *c.*1430. See THRUSTLE (under THRUSSEL), also THRICE COCK.

Thrusher, see THRESHER.

Thrushfield. A Shrops. expression for the Thrush, which we take to be a conflation of THRUSH and FIELDFARE.

Thrussel, etymological spelling **Thrustle.** A variant of THROSTLE recorded from Shrops., Yorks., and Scotland, due to contamination from THRUSH. It is found as early as *c.*1350 'thrustele'.

Thumb Bird. A Hants. name for the Goldcrest, an allusion to the bird's smallness, as in MILLER'S THUMB.

Tiddy Wren, also **Tidley Wren.** Local Essex names for the Wren, the epithets meaning tiny; cf. TITTY WREN.

Tidley Goldfinch. A Devon name for the Goldcrest, *tidley* meaning tiny.

Tink. A name in Cornwall and Devon for the Chaffinch, possibly Celtic, but in any case onomatopoeic. For similar patterning, see PINK.

Tinkershire. A Dorset name for a Guillemot, first recorded in 1799. It is basically **Tinker**, a seaman's pet name, evasively extended to give the present, nonsensical form. Cf. SEA HEN.

Tinnock. A name for the Blue Tit, locality not specified. Formed with the suffix -OCK from a syllable *tinn-* echoing the shrill, trilling song. Cf. PICKATEE.

Tirma. An Oystercatcher name known from Martin 1698 'Tirma, or Sea-Pie, called by the inhabitants *Trilichan'*. In this Scottish name, *tir-* is plausibly an onomatopoeic representation of the cry, *-ma* a form of MAW. The bird has a considerable vocal range; cf. CHALDER, OLIVE.

Tit. In origin a colloquial abbreviation of TITMOUSE, making its debut in a dictionary entry of 1706 'tit, or Titmouse', As the quotation implies, the short form competed with the full form, at first of course in the spoken language, then from the beginning of the 19th cent. sporadically in the written language as well, until Yarrell 1843 generalized the abbreviated form. Nevertheless much hesitancy remained, and Titmouse persisted for long as the commoner expression in ornithological work, as in BOU 1883, Hartert 1912, even *Checklist* 1952. But since about 1930, Tit may well be regarded as the *de facto* normal form.

Tit Lark. The best known traditional name for a Pipit, first in Merrett 1667. The element TIT denotes a small creature; see TITMOUSE. The literal meaning of the name is thus Little Lark, a Pipit being noticeably smaller than the otherwise not dissimilar Lark, i.e. the Skylark. Cf. CHIT LARK. Pipit names based on LARK may involve an onomatopoeic element, as Ches. **Peet Lark**.

The present name was used by some naturalists in the earlier part of the 19th cent., until Yarrell 1843 condemned it on the grounds that it was popularly applied indiscriminately to all three species of Pipit found in Britain.

Titling. A north country and Scottish name for the Meadow Pipit, the record going back to the mid-16th cent. The word derives from Old Norse *titlingr* pipit from original **títlingr*, i.e. with regular shortening of the root vowel

before *tl*, where *tít-* is imitative of the Pipit's sharp call-note, *-lingr* a derivative suffix, see -LING. Cf. TEETING.

Titmouse. First appearing *c.*1325 'titemose', next *c.*1400 'titmose', the name is a compound consisting of Middle English *tit*, a general term for any small creature, and *mōse*, the earlier name for these birds, last recorded *c.*1250. The latter regularly continues Old English *māse*, of Germanic age, paralleled in Dutch *mees* and German *Meise*; Old Norse has *meisingr* (the suffix *-ingr* having a diminutive function). The basic sense is small creature, in fact semantically equivalent to *tit* (above). Cf. OX EYE.

The form *titmōse* last appeared in 1570, but meanwhile in 1530 the now current form had made its debut and quickly became general, *-mouse* replacing *-mōse*, which was no longer understood. Not improbably, the mouse-like movements of these birds encouraged the corruption, with plural *-mice* following as a matter of course.

Generically equated since Ray 1678 with *Parus*, the present term eventually gave way, to a large extent, to the abbreviated form TIT, as explained under that name. The standard specific terms involving Titmouse, as BLUE TITMOUSE, are not genuine folk-names, but translations of Latin prototypes.

Cf. COALMOUSE, CREAK MOUSE, also STRAW MOUSE.

Titterel. A Whimbrel name reported from Sussex and Norf., imitative of the distinctive call, and formed from the echoic verb *titter* under the influence of the suffix -EREL.

Titter Wren. An Essex variant of TITTY WREN, i.e. with substitution of -ER for -Y, as a formation comparable to GILLER WREN.

Titty Mouse. A not unexpected hypocoristic extension of TITMOUSE. First recorded in the 16th cent., it survives locally in Somerset in the specific sense Blue Tit.

Titty Todger. A Devon name for the Wren. For *titty*, see TITTY WREN, but a precise interpretation of *todger* is difficult. However, the -ER suffix is common in Wren names, see GILLER WREN, and cf. STUMPY TODDY.

Titty Wren. A popular name for the Wren, found in various places from Suff. to Devon. The epithet derives from *tit* small creature; see TITMOUSE. Cf. TITTER WREN, also TIDDY WREN, CHITTY WREN.

Tod Bird. A name for the Great Skua in use on the Yorks. coast, *tod* reproducing the local pronunciation of turd (Old English *tord*). Motivation as in DUNG BIRD.

Tom Harry. A name in Cornwall for the Great Skua, apparently so called from its harrying pursuit of other seabirds. There are variants **Tom Horry**, the latter element formally a dialect word for dirty, and **Tom Hurry**. Which of these three will be the primary name is anybody's guess. It must also be borne in mind that the second element may well be the original name, to which *Tom* was added later, as is usual when a Christian name occurs as the first element. And all this only on the assumption that the present forms are not hopeless corruptions of a forgotten original which, in this area, would most probably be Celtic.

Tom Thumb, see MILLER'S THUMB.

Tom Tit. In general use as a pet name for the Blue Tit, known since 1709.

Tope. A name for the Wren from Devon and Cornwall. Typological considerations indicate that an English vernacular name for the Wren will refer either to the smallness of the bird or else to its cocked-up tail, see further under WREN. The present name, however, is not at first sight ascribable to either of these.

We next note that *tope* is the standard name of a small shark *Galeorhinus galeus*, hitherto assumed to be different in origin from the bird name, though neither could be explained. The identification of *tope* with *G. galeus* is due to Ray, *Historia Piscium*, 1686, but with an element of doubt in the matter, as the naturalist's words show: 'a Cornubiensibus, ni fallor, *A tope* dicitur' i.e. 'it is called by Cornishmen *a tope*—unless I am mistaken.' The shark name further occurs in a variant form *toper* (with redundant -*er*, as sometimes in bird names, see -ER), cf. Nance 1963 '*toper* . . . the porbeagle (at) Porthleven.' In Norfolk the name is given to the spur dog-fish, *Squalus acanthias.*' Can we now deduce the primary meaning of *tope/toper*? For the sake of argument, we shall assume it to be the last mentioned. The species in question has a poisonous spine (the spur) in front of each of the dorsal fins and, almost invariably, takes its name from this dangerous weaponry. For example, in Old Norse the creature is called *hár*, a word also having more literal meanings, as pin, peg, small stick. We may suppose that the same could apply to *tope* and see our supposition confirmed in the Wren name, since this bird's names are commonly motivated by the unique tail, here compared to a small stick or the like. The shark and bird names are thus as one, preserving in figurative usage a word forgotten in its original sense.

Tor Ouzel, see ROCK OUZEL.

Tot o'er Seas, see HERRING SPINK.

Tree Creeper. Formerly CREEPER, in reference to the characteristic movements, the bird creeping, mouse-like, over the trunk and boughs of trees, as it searches for the minute life upon which it feeds. There are similar allusions in local names, as Wilts. **Tree Climber**, Oxon. **Tree Clipper**, or Scottish **Tree Speeler** (*clipper*, *speeler* being local words for *climber*). The word *bark* is sometimes substituted for *tree*; see BARK CREEPER. Cf. TREE MOUSE.

The present name does not appear in the records until 1814. It replaced COMMON CREEPER as the standard name for the species in BOU 1883.

Tree Goose. A term for the Barnacle Goose, first noticed in 1597, but doubtless much older, as it appears, translated, in (medieval) German *boumgans* in Albertus Magnus *c*.1250. Albertus refutes a Scottish legend to the effect that the fruits of certain trees fall into the sea, there to develop into Geese; see further under BARNACLE GOOSE.

Tree Mouse. A Somerset name for the Tree Creeper, motivated by the mouselike creeping movements.

Tree Pipit, see PIPIT.

Tree Sparrow. This species, formerly known as the MOUNTAIN SPARROW, was not, it appears, specifically distinguished from the House Sparrow until 1713. The present term was coined by Pennant 1770 'Tree Sparrow. Mountain Sparrow. Common near Lincoln . . . conversant among trees, but does not frequent houses'; it became general in the early 19th cent.

Tree Speeler, see TREE CREEPER.

Trillachan. Given by Martin 1698 as a St. Kilda name for the Oystercatcher. Elsewhere in the Hebrides this word is used of other birds, as the Sandpiper or Sanderling. The reason for these shifts of meaning lies in the name itself which, in Gaelic, simply means triller; it consists of the onomatopoeic root *trill* with the two suffixes *-ach* and *-an*; cf. PTARMIGAN. In a Gaelic text, the name is properly spelt *trilleachan* or *triollachan*, in accordance with the spelling conventions of that language (and the phonetic nature of *ll*).

Tufted Duck. First in Ray 1678, *tufted* translating Latin *cirratus* found in *Mergus* (diver) *cirratus*, the name coined by Gesner 1555. The epithet refers, of course, to the tuft of feathers hanging down from the back of the head. The name was adopted by Pennant 1768 and used by virtually all his successors.

Tuke. A Redshank name from Essex, imitating the alarm note. See CLEE.

Tuliac. A spurious form, stalking through the literature since Macaulay 1764 supplied it as the St. Kilda term for the Great Skua. But, as Pennant 1768 observed, the description given was rather that of the Herring Gull, and this removes any doubt that Tuliac is a scribal or printing error for **Fuliac,** i.e. Gaelic *faoileag*. Otherwise the present form is a fair representation of the actual pronunciation of the Gaelic in terms of English spelling. It is noticeable that Macaulay makes the same mistake in reporting FULMAR as **Tulmer.** A comparable corruption is seen elsewhere in FEASER.

The Gaelic word shows the diminutive ending *-ag*. Other dialects may know this word as *faoilean*, i.e. with the *-an* suffix (as in SOLAN), and this is the older form of the name which goes back to Old Irish *foilenn*. This last, originally **oilenn* (secondary *f* before initial vowel being a common feature of Gaelic, cf. FALK), was borrowed from Late British Celtic **uilan*. It can be demonstrated philologically that this borrowing could not have been earlier than 600, most probably taking place between then and 750. The present name is thus ultimately of the same origin as GULL.

Tullet. A name for the Ringed Plover from Ches. and Lancs., imitative of the musical call. Cf. DULWILLY.

Tulmer, see TULIAC.

Turnstone. Ray first published the name in 1674; in 1678 he put 'Turn-stone, or Sea-Dotterel'. Pennant 1768 opted for 'Turnstone', effectively standardizing it. Where the name actually arose is not known, but the motivation is clear: the birds congregate on stony beaches, moving aside stones, pebbles, indeed any small object, as they search for the marine life on which they feed. The same imagery is found in STANE PUTTER. Seaweed, too, is inspected and moved, cf. TANGLE PICKER.

Turtle. An archaic name for the Turtle Dove, going back to Old English *turtle*, *turtla*, dissimilated from **turture, turtura*. The term was taken from Latin *turtur*, familiar through its occurring in the Latin Bible (Lev. 14:22, Luke 2:24). The name was commonly used throughout the Middle Ages. It is found in Caxton and Shakespeare, but afterwards became unusual. Surprisingly, Pennant 1768 has it, but in subsequent editions we find TURTLE DOVE. The word is naturally onomatopoeic.

Turtle Dove. First recorded *c*.1300, DOVE having been added to the already existing TURTLE. Turner 1544 uses the present form and most naturalists followed suit.

Twink. A local name for the Chaffinch in use from Devon to Ches.; it is clearly echoic, see comparable synonyms under PINK.

Twite. First apparently in 1562 'twyes' (considered to be a misprint for 'twytes'), next 1676 'Twite'. It occurs in Albin 1738 and was taken up by Pennant 1776, but MOUNTAIN LINNET was generally used by his successors, including Yarrell 1843 and later editions; Yarrell, however, quoted Twite as a popular name. Newton 1896 regarded both names as permissible, though Twite had by then become the more usual, having appeared in BOU 1883. It figures in Hartert 1912, since when it may be definitely described as the recognized standard.

Twite is echoic, and was felt to be so by the bird-catchers who used the name, but the original form was more likely **Tweet*, a closer representation of the call. In this case, the present form will have resulted from sound changes taking place in the language at large, the diphthongization envisaged taking place in the 15th cent., so that the name will be at least as old as that. It is said to have been used in the countryside near London and made known in the capital through the trade in caged birds.

The layman does not readily distinguish this species as such; if he does, he regards it as a type of Linnet, see HILL LINTIE.

Two Fingers. A Ches. name for the Wren, expressing the concept of smallness.

Tyst, now **Tystie.** A term for the Black Guillemot, native to Shetland, Orkney, and Caithness, the latter form with added diminutive -IE; the first records are 1744 'toists' and 1806 'teisty' respectively. This is a Norn word, the source being Old Norse *þeisti*, a name imitative of the distinctive, soft whistling call.

Tystie has occasionally been taken up by recent writers, e.g. Fisher 1966 'Black Guillemot or Tystie', since the species (formerly *Uria grylle*, now *Cepphus grylle*) is today generally separated from the Common Guillemot (*Uria aalge*). In these circumstances, the standard term BLACK GUILLEMOT may be felt rather less appropriate.

U

Utick. A name for the Whinchat recorded in many localities from Middlesex and Oxon. to Cumb. and Yorks. It is based on the distinctive call note, heard as 'Utick-tick-tick'. Cf. EAR TICK.

V

Vare Wigeon. A North Devon name for the Smew, *vare* being a local word for a weasel, motivation as in WEASEL DUCK.

Veldiver, Velly Bird, Velverd, see FIELDFARE.

Velvet Duck. A designation for the Velvet Scoter proposed by Ray 1678 'The feathers of the whole body are so soft and delicate as nothing more, so that it might be not undeservedly called the Velvet-Duck.' Pennant 1768 took up this name, which remained in use until Fleming 1828 introduced the term employed today.

Velvet Scoter. Coined by Fleming 1828 for systematic reasons, replacing VELVET DUCK.

W

Wagel. An obsolete name, all citations of which derive ultimately from Ray 1678 'Wagell' who defines the term as 'the great gray Gull', apparently an immature form of the Great Black-backed Gull, formerly regarded as a separate species. The term, taken down near St. Ives *a*.1672, is doubtless a relic of the Cornish language. It cannot be etymologized, but the radical form would be **gwagel*.

Wagstart. Recorded between *c*.1440 'wagstert' and 1570 'wagstarte'. In spite of the late attestation, we have here an old word for the Wagtail, to all appearances of West Germanic age, as strongly suggested by the Middle Low German synonym *wagestert*. See -START, WAGTAIL.

Wagtail. First in 1510 'wag tayle', and soon to replace the older name WAGSTART; it was used by the early naturalists. The term is, of course, indifferently used of all the Wagtails, and the same may apply to the common synonym WATER WAGTAIL; see also WASHTAIL. The three native Wagtails are officially distinguished by epithets alluding to the colouring of their plumage: PIED WAGTAIL, formerly WHITE WAGTAIL, GREY WAGTAIL, YELLOW WAGTAIL.

Wall Bird. The Spotted Flycatcher often sites its nest against a creeper-covered or trellised wall, hence the present name found from Norf. to Hants. In the North Riding of Yorks. the corresponding name is **Wall Chat** (see CHAT), in Ches. **Wall Robin**, its usual call being reminiscent of the Robin. Cf. BEAM BIRD.

Warbler. A generic term introduced by Pennant 1773 to render Scopoli 1769 *Sylvia* literally (bird) belonging to the woodlands. Evidently both words were chosen for want of something better.
It will be noticed that several Warblers have since been removed from the genus *Sylvia* to various other genera; those referred to in this book are the Dartford Warbler, the Grasshopper Warbler, the Willow and Wood Warblers, and finally the Marsh, Reed, and Sedge Warblers, their vernacular names thus witnesses to the systematics of an age gone by.

War Cock. A name for the Water Rail (Jackson 1968, no locality given), to all appearances regularly continuing Middle English *c*.1420 *wercock* (species not indicated). In the light of this, the Wilts. variant **Oar Cock** is evidently corrupt. The proper sense of the first element has not been determined. *Cock* is a constituent of four other Water Rail names, in three of which the first element refers to the characteristic running: BILCOCK, DARCOCK, SKIDDY COCK, the fourth being GUTTER COCK.

Ware Goose. A term for the Brent Goose from the Durham and Northumb. coasts, so called from its feeding habits, *ware* being a traditional word for seaweed.

Wariangle. An obsolete Shrike name, first *c*.1386 'waryangles', next in 1598 when 'warriangles' was said to be a common name in Staffs. and Shrops.; it appears in Cotgrave 1611 'wariangle' and in more recent times in Ray 1674 'the great Butcher-bird called in the Peak of Derbyshire Wirrangle' and Ray 1678 'This bird in the North of England is called a Wierangle.' Somewhat different from the foregoing is a variant from 1618 'Warwinckle'.
Although not encountered in the records of Old English, the name is nevertheless of West Germanic age, as shown by the Old High German cognate *wargengil*, surviving locally as *Wörgengel*. It is, however, not possible to construct a prototype with any confidence, so that the etymology of this,

our oldest Shrike name, remains elusive.

Washtail. The Wagtail, so often seen by streams and ponds, has been nicknamed Washtail, also **Nanny Washtail**, the constant up-and-down movement of the tail suggesting the battledore once used to beat the washing. From such a name, it was only a step to other familiar forms as **Moll Washer** or **Washerwoman**. Elsewhere the sense shifted to washing-up, hence **Dishwasher** or **Washdish** which, with their self-evident synonyms **Peggy Dishwasher** or **Polly Washdish**, are also known from many parts of the country. It may be added that all these names refer in practice chiefly to the Pied Wagtail as by far the commonest of the native species.

The nicknames above are paralleled in local French *batte-lessive* literally beat-(the)-washing and *lavandière* literally washerwoman, recorded by Belon 1555. Such names are not found in German, so that it is very possible that French influence ultimately accounts for the English tradition.

Water Blackbird. A Scottish name for the Dipper.

Water Crake. A northern name for the Dipper, first in Ray 1768 'Water Ouzel, or Water Crake', an alternative to WATER CROW; see CRAKE 2.

Water Crow. A northern and Scottish name for the Dipper, first in Turner 1544 'water craw'. Cf. WATER CRAKE.

Water Hen. The record of this well-known name goes back to Skelton *c.*1508; it is given as a common alternative to MOORHEN by many early naturalists, beginning with Ray 1678 'Waterhen, or More-hen'. Pennant 1768 called the bird 'Common Water Hen'; see GALLINULE.

Water Ouzel. A Dipper name from Somerset, Glos., Worcs., Warwicks., and Yorks., known since Ray 1678 'Water Ouzel, or Water Crake'. It was

used by Pennant 1768 and many of his successors until replaced in the standard nomenclature by DIPPER. See BROOK OUZEL.

Water Piet. A Scottish name for the Dipper, PIET i.e. magpie, being suggested by the stark contrast between the white of the breast and the dark hues of the rest of the plumage.

Water Rail. First in Mouffet 1655 'water-rails', next in Ray 1678 'The Water-Rail, called by some the Bilcock or Brook-Ouzel'. The name is not traditional, but a translation of Aldrovandi 1603 *Rallus aquaticus*. This book name was adopted by Pennant 1768 and met with the approval of his successors. See LAND RAIL.

Water Wagtail. Often found as a synonym of generic WAGTAIL, doubtless traditional, from the commonly preferred habitat.

Waxwing. The periodic eruptions of this Arctic bird reach Central Europe as well as the British Isles. Just as we in this country sometimes employ the epithet *French* to characterize a less usual, migratory species, so the Germans may use *Böhme* (a) Bohemian, and this they do in the case of the Waxwing. Such popular usage served as a model for Gesner 1555 *Garrulus bohemicus* literally Bohemian Jay. In the absence of a native English name, Ray 1678 adopted this term, but mistranslated it as **Bohemian Chatterer**, especially unfortunate seeing that the Waxwing is a very silent bird. Edwards 1758, describing the bird from Catesby 1731 as 'the Chatterer of Carolina', drew attention to 'those singular scarlet wax like appendages to some of the feathers of the wings', an observation which led to Pennant 1776 'Waxen Chatterer', the term adopted by his immediate successors. But after Vieillot, in 1816, had grouped the Waxwings into a genus of their own, to be known as *Bombycilla* (and separate from the Chatterers), Stephens 1817 coined Waxwing for use as the vernacular counterpart of *Bombycilla*, a convenient term which gained immediate acceptance. See SILK TAIL.

Weasel Duck. A name for the Smew used in Norf. and Suff., also in Northumb., inspired by the appearance of the female bird, her white throat and cheeks contrasting, weasel-like, with the chestnut-brown feathers of the head and nape. The same imagery is seen again in the synonym VARE WIGEON.

Weeo. An Orkney name for the Kittiwake, an onomatopoeic word (Norn *wee* squeal), most appropriate for this screaming Gull. It corresponds formally to Old Norse *-vía* seen in *langvía* guillemot (beside *langvíi*, *see* LONGIE).

Weerit. A Renfrew name for a young Guillemot; it is an imitation of the juvenile call, see further under GUILLEMOT.

Weirangle, see WARIANGLE.

Wet Bird. A Somerset term for the Green Woodpecker, motivated by the once common belief that its call portended rain, cf. RAIN BIRD, see also WHETILE.

Wet-me-lip. A Norf. name for the Quail, in imitation of the loud call. For another interpretation, see QUICK-ME-DICK.

Whaup. The common northern and Scottish name for the Curlew. Although not actually attested until 1538, it is nevertheless traditional and evidently presupposes (Northern) Old English *hwalp* to be compared with the recorded Old English sea-bird name *hwilpe*. The precise meaning of the latter cannot be determined, but it is akin to Middle Low German *wülpe* (contained in the place name *Wülpensand*), the approximate sense of which can be deduced from the modern word *Regenwölp* (*Regen* rain), a Mecklenburg term for a Curlew or a Plover, further Dutch *wulp* curlew. These are imitative words, applicable to various species, indeed also to other animals, cf. English *whelp* puppy, and see CURLEW HELP.

Wheatear. To all appearances first attested in 1591 'thrie whekeres', presumably a misprint for *'wheteres'*, next 1653 'Th'are called Wheat-ears', then *a.*1661 'Wheat-ears is a bird peculiar to (Sussex) . . . It is so called, because fattest when wheat is ripe . . . whereon it feeds.' Ray 1678 has 'Wheat-ear, Fallow-smich, White-tail'; from these Pennant 1768 took 'Wheatear' which thereafter was used as the standard name. Ray, *op. cit.*, characterizes it as a Sussex name; cf. EAR BIRD.

The element *wheat* is obviously the product of popular etymology, as seen in the explanation of the name of this insect-eater *a.*1661. The same applies to *ear*, but at the same time the use of 'wheat-ears' as a singular will have been original, the name surely standing for Middle English *whiters* literally white arse; cf. the local synonyms WHITE ASS, WHITE RUMP. The word *arse* did not sink to vulgar status until the 17th cent., cf. ARSE FOOT.

The Wheatear has many local names, often imitative, the onomatopoeic element being commonly preceded by *fallow* or especially by *stone*, see under FALLOW CHAT, STONE CHACK.

Wheel Bird. A name from Stirling for the Nightjar, the characteristic burring song suggesting the whirring of a spinning wheel.

Wheetie. A Scottish term for a Whitethroat, based on the echoic syllable *wheet-* in imitation of the song; the ending is the usual diminutive -IE.

Whetile. A name for the Green Woodpecker recorded from Herts., Worcs., and Essex. As the Herts. pronunciation /wetoil/ confirms, this name was originally **Wet Tail**, of which Somerset **Whittle** is a further corruption. The name was inspired by the superstition that the bird's call presaged rain; cf. RAIN BIRD, WET BIRD. See also GREENILE.

Whew. A Northumb. word for the Wigeon, imitative of the drake's whistling call, in Wilts. **Whew Duck,** in

Norf. **Pandle Whew**, pandles being winkles taken with the bottom-growing vegetation upon which this Duck mainly feeds. Further **Whewer** literally whistler from Ray 1678 who adds 'The Males . . . in *Cambridge* are called *Wigeons*, the Females *Whewers*', obviously a purely local narrowing of the sense.

Whimbrel. First in Ray 1678, the term being supplied by a correspondent in the North Riding of Yorks. It was taken up by Pennant 1768, who thus effectively gave it standard status. The name is basically imitative of the call represented by northern dialect *whimmer* whimper, influenced by the suffix -EREL, i.e. originally *whimmerel*, contracting to *whimrel*, then developing an excrescent *b* (as in *embers* from Middle English *emers*), whence the recorded form.

A related **Whimperel**, based on *whimper*, is attested in 'Whymperel' from a Durham source of 1530/31. For another close northern relative, see WINDEREL, the influence of the suffix -EREL again in the synonyms CHICKEREL, TITTEREL.

Other names for this species include BREEM, HALF CURLEW, LITTLE WHAUP, SEVEN WHISTLER, SPOWE, TANG WHAUP.

Whinchat. First in Ray 1678 'whin-chat'; it was adopted by Pennant 1768 and so effectively standardized. This bird nests in rough vegetation, often at the foot of a whin bush. So does the Stonechat, to which species the present name may be applied locally. In fact, popular usage frequently fails to distinguish between the two at all, since they are so very much alike in appearance and habits. The choice of Whinchat to indicate *Saxicola rubetra* rather than *S. torquata* was thus, in the last analysis, fortuitous. See FURZE CHAT, GORSE CHAT.

For a distinctive Whinchat name, see UTICK.

Whindle, see WHIN THRUSH.

Whin Thrush. A Glos. term for the Redwing. This species may receive its name from a unique, whining note, and

we postulate Middle English *whinthrushe* literally whine thrush. In accordance with the phonetic tendencies in the language, long vowels were later generally shortened when they occurred before a consonant cluster, hence original *whin-* became *whin-*.

The first syllable was sometimes corrupted, as in **Wing Thrush**, known from Turner 1544 'wyngthrushe', or Devon **Wind Throstle** attested in 1674 'Wind-throstle, or Whindle', also **Wind Thrush**, first found in Charleton 1668 and surviving in Warwicks., Glos., Somerset, and Cornwall.

Evidence from Cornwall and Devon shows that the onomatopoeic syllable could, of itself, form the basis of a name, hence with the suffix -ARD we have the Cornwall name **Winnard**, recorded in 1758, earlier 'Wheenard' from *a*.1698, further Devon **Whindle** (above), now **Windle**, reflecting archaic dialect *whindle*, today *windle* whine. See SWINE PIPE.

Whiskered Tern. The bird was identified as a British species in the 1830s and Gould 1837 coined the name **Moustached Tern**, which Yarrell 1843 changed to Whiskered Tern, the moustache or whisker being the strip of white running from the gape to the ear-coverts.

Whistling Plover. A term from various parts of Great Britain denoting either the Grey or the Golden Plover, both species having a shrill, whistling call and being otherwise quite similar. The name is first encountered in Merrett 1667.

Whistling Swan. A local Northumb. term, introduced by Pennant 1785 in place of WILD SWAN. Some of his successors accepted this innovation, but Yarrell 1843 chose 'Hooper', i.e. WHOOPER, as the main name, after which the present term passed out of vogue.

White Ass. A name in Cornwall for the Wheatear, properly **White Arse**; cf. WHITE RUMP and see WHEATEAR.

White-fronted Goose. Coined by Pennant 1768 in preference to the folk-name LAUGHING GOOSE, and general ever since.

White Lorin, see LORIN.

White Nun. A name for the Smew introduced by Ray 1678 'We may call it with the Germans the White Nun', his inspiration being L. Baldner, *Vogelbuch*, 1666 'ein grosse weisse Nunn'. In Germany, this and similar terms are traditional, going back to the 15th cent. (Suolahti 1909). The fillet of black markings on the head of the male, set off by the surrounding white, recall the appearance of a nun; cf. our own Blue Tit name NUN.

White Owl. A widespread name for the Barn Owl, so called from the white underparts. First found in Ray 1678, it was employed by Pennant 1768 and some of his successors, partly as an alternative to BARN OWL, as explained under this name. Cf. YELLOW OWL.

White Puffin. A Manx name for the Fulmar, presupposing Manx Gaelic *pibbin vane* literally white shearwater. The name must have arisen before the Fulmar (and its name) became so well known. Until the last quarter of the 19th cent., the Fulmar was still a rare bird in Manx waters, so that when the Manxmen saw it, they referred to it in terms of the Shearwater, so familiar as the 'Puffin of the Isle of Man'. *JMM* vii 118f. (1965)

White Rump. A name for the Wheatear from Cumb. and Northumb., also from Norf. The word *rump* is a Scandinavian loan, first recorded in English in *Prompt. Parv.* c.1440 'Rumpe, tayle, cauda'. Cf. WHITE ASS, also WHEATEAR.

Whitestart. Literally white tail; see START. A Yorks. (Wharfedale) word for the Wheatear. Cf. WHITE TAIL.

White Stork. Introduced by Ray 1678 'Common or white stork', a self-explanatory specific term which auto-matically became part and parcel of our ornithological vocabulary.

White Tail. A term for the Wheatear, known to have been in use in Sussex, first found in Cotgrave 1611 'whittaile *culblanc*'; it was quoted by Ray 1678 'Wheat-ear, Fallow-smich, White-tail'. The same name appears in Cornwall as **Wittol**; such a form presupposes Middle English **whíttayl*, i.e. with regular shortening of the long vowel before the double consonant. See TAIL and cf. WHITESTART.

White-tailed Eagle. A name due to Edwards 1743, inspired by Gesner 1555 *Albicilla*, a term coined by Gaza 1476 from Latin *albi-* white and *-cilla* tail to translate Greek *pýgargos* literally *pýg*-rump, *-argos* white. (Gaza's *-cilla* tail is spurious; he extracted it from *mōtacilla* wagtail, interpreted as *mōta-* move and *-cilla* therefore tail. In reality, however, the name is to be analysed *mōtac-* —reminiscent of Greek *múttēx*, stem *múttek-*, species not known—and *-illa*, the common diminutive ending.) In Ray 1678 we read 'white-tail'd Eagle, called *Pygargus*, and *Albicilla*', in Pennant 1768 'Pygargus, or white-tailed Eagle. Ern', but in 1776 this was replaced by CINEREOUS EAGLE. Bewick 1804, however, took up the present name and Yarrell 1843 effectively standardized it, though SEA EAGLE still occurs as an unofficial synonym; see also ERNE.

Whitethroat. Though not attested until 1676, the name is surely considerably older. As a name type it is comparable to REDBREAST and will have arisen about the same time, say in the 14th cent. It occurs in Ray 1678 and Pennant 1768 and nearly all other writers, so that it has been the standard name as far back as the record goes. See LESSER WHITETHROAT.

White Wagtail. A name for the Pied Wagtail first noticed in Ray 1678, doubtless a translation of Gesner 1555 *Motacilla alba*. Though taken up by Pennant

1768, his successors generally preferred PIED WAGTAIL. The present term, however, later came in handy to distinguish that very closely related summer visitor from the Continent, Linnaeus 1758 *M. alba*, whose migration to this country was being postulated by the 1830s.

White Whisky John. A name for the Great Grey Shrike, locality not given. *White* refers to the colour of the underparts, while *whisky* alludes to the characteristic whisking, i.e. sweeping, of the tail from side to side; cf. FLUSHER.

White Wigeon. A Devon name for the Smew.

Whittle, see WHETILE.

Whooper, officially **Whooper Swan.** Formerly, and correctly, spelt Hooper, a name motivated by the powerful, hooping call. Known since 1566 'A hooper or wilde swanne', it was quoted by Ray 1678 'A wild Swan, called also an Elk, and in some places a Hooper'. Pennant 1768 took 'Wild Swan' as his standard term, altering this in 1785 to 'Whistling Swan' and succeeding authorities used one or other of these until Yarrell 1843 chose 'Hooper' as the chief name. As usual, Yarrell's precept was widely accepted. The present spelling, a purely orthographic change, became general in the last quarter of the 19th cent. At the same time, the name was often expanded to Whooper Swan, thus bringing the name stylistically into line with the other standard Swan names: BEWICK'S SWAN, MUTE SWAN, as in BOU 1883.

Wigeon, obsolescent spelling **Widgeon.** The name of this whistling Duck makes its debut in 1513 'wegyons'. It was used by the early naturalists, e.g. Turner 1544 'Wigene' Latinized plural, Merrett 1667 'Widgeon', Ray 1678 'Wigeon or Whewer'. Pennant 1768 chose 'Wigeon' which thus became *de facto* the standard expression. It is a borrowing of synonymous Old French *vigeon* which goes

back via *vibiōnem* of the Late Latin of Gaul to Classical *vipiōnem*, accusative of *vipiō*, of onomatopoeic origin. For parallel phonetic changes, cf. PIGEON.

Other names for this species are, as in the present case, typically imitative of the drake's far-reaching, whistling call, cf. the native English terms WHEW and WHEWER, also WINDER; see further under SMEW.

Wild Duck. A once widely used name for the Mallard which is first noticed in Merrett 1667 '*Anas fera* Wild Duck', and which remained common in the work of the naturalists, generally as the main term, until the beginning of this century; it was used in BOU 1883, see further under MALLARD. At the same time, the present expression could be restricted to the female of the species, the male then being called the Mallard. Since Mallard in this sense goes back to the early 14th cent., it seems to follow that Wild Duck is also at least as old as that.

Wild Swan. To early observers—they recognized only two species of Swan— the Whooper, the winter visitor, contrasted with the Mute, familiar as a semi-domesticated species, according to a tradition since the days of Richard Lionheart. For them the Whooper was the Wild Swan, the Mute the **Tame Swan**, and these expressions were chosen by Pennant 1768. But on learning later that the Mute Swan nested wild in Russia, Pennant 1785 substituted WHISTLING SWAN and MUTE SWAN respectively; see further under these names. Nevertheless, the present term continued to be used by several authors until Yarrell 1843 established the use of 'Hooper', i.e. WHOOPER, as the standard term, though he, too, uses Wild Swan as a stylistic alternative.

Wilkie, Will, Willick, Willock, see GUILLEMOT.

Willow Biter. A Notts. term for the Marsh or Willow Tit, species which may nest in willows. For *biter*, cf. BILLY BITER.

Willow Tit, in older style **Willow Titmouse.** This bird, for so long not distinguished from the Marsh Tit, was shown to be a native British species in 1897. The term itself was suggested by the scientific name *Salicarius*, based on Latin *salix* willow.

Willow Warbler. Coined by Yarrell 1843 for systematic symmetry in place of the earlier WILLOW WREN.

The Willow Warbler constructs a dome-shaped nest, well hidden on the ground, one which has some similarity with the remarkable nest of the Long-tailed Tit. Accordingly, names for this Tit, in so far as they allude to the nest (see LONG-TAILED TIT), may be transferred locally to the present bird.

Willow Wren. A name occurring locally, mainly in the south, first in Pennant 1768 and used by some of his successors until Yarrell 1843 substituted WILLOW WARBLER.

Willy, see GUILLEMOT.

Willy Wicket. A Sandpiper name, common in the north of England, in the last analysis echoing the shrill cry.

Wind. A south country name for the Dotterel, which we take to be basically imitative of the soft, fluting call, perhaps originally *Whind, in which connection compare WINDER, WINDEREL.

Wind Bibber. A name for the Kestrel from Sussex and Kent. The expression could be taken as meaning wind drinker (*bib* to drink), but it is based on the local word *bibber* shake, tremble, in allusion to the hovering bird.

Wind Cuffer. An Orkney term for the Kestrel, *cuffer* suggesting, however inadequately, the beating wings of this hovering hawk, cf. WIND FANNER. At the same time, the unidiomatic use of *cuffer* is in itself an indication that this part of the name cannot be original. It is pretty clearly a euphemistic transposition of the consonants seen in *fucker,* the

primary form, discussed under WIND FUCKER.

Winder. A Kent name for the Wigeon. Since the names of this species are commonly derived from the characteristic call note, see further under WIGEON, the present name is to be connected with the verb *winder* (*wind-* pronounced as in the *wind*), originally *whinder, a variant of *whimper* with the same meaning. In Norf. the bird may be known locally as the **Cock Winder.**

Winderel. A Northumb. name for the Whimbrel, originally *Whinderel, based on the verb *whinder* to whimper (see WINDER) and having the distinctive ending -EREL, so characteristic of onomatopoeic Whimbrel names. It is thus closely related to Durham **Whimperel,** and to WHIMBREL itself, see under this name.

Wind Fanner. A Sussex and Surrey term for the Kestrel, motivated by the vigorously flapping wings of the hovering bird. Cf. FANNER HAWK.

Wind Fucker. An obsolete term, known from a single record of 1599 'The kistrilles or windfuckers that fill themselues with winde, fly against the winde euermore' (*OED*). It may be compared with another Kestrel name **Fuckwind,** given as a northern term by Halliwell 1847.

The *OED* also records the present word as an expression of opprobrium in sources from the early years of the 17th cent. e.g. 1609 'Did you euer heare such a Wind-fucker, as this?' In a text *a*.1616 (ed. 1639) the word has been altered to 'windsucker', evidently a euphemism, which also occurs as a Kestrel name, see WIND SUCKER. For a similar euphemism, cf. WIND CUFFER.

To interpret the names Wind Fucker and Fuckwind we need to establish if *fuck*, attested since 1503, could once have had another meaning. We notice the cognate Dutch *fokken* to fuck, also to knock, then German *ficken* to fuck, in dialect to beat. Taking the sense of knock, beat to be more original, and assuming the same for English, we at once have an explanation of the bird names: they are literally wind beater and beatwind, in allusion to the beating wings of the hovering bird, as in the synonym WIND FANNER.

The present bird names thus uniquely preserve *fuck* in its hitherto unrecognized, primary sense. Needless to say, the names in question must have come into existence before the word acquired its current meaning, i.e. before 1503 at the latest. It follows that the opprobrious use, referred to above, is a vulgar application of the bird name, arising after the original sense had been forgotten.

Windhover. A term for the Kestrel, common in southern and western counties, and often used in literature. It was first recorded by Ray 1674 'The Kestrell or Stannel, in some places the Windover', also in 1678 in the text 'Windhover'. The literal sense is wind hoverer, the element -*hover* being formed from the older verb *hove* to hover, which suggests that the name goes back to Middle English times.

Other Kestrel names alluding to the characteristic, hovering flight include WIND BIBBER, WIND FANNER, and WIND FUCKER with its derivatives.

Windle, see WHIN THRUSH.

Windle Straw. A Shrops. name for a Whitethroat, *windle* meaning stalk of withered grass. The term thus alludes to the nesting material, the name type paralleled in SMALL STRAW.

Window Swallow. A Northumb. name for the House Martin. The bird would often site its nest under the window-sill or in some niche around the old-fashioned windows. First in Bewick 1797 'The Martin, Martlet, Martinet, or Window-swallow'.

Wind Sucker. A Kestrel name from Kent, modified by popular etymology from WIND FUCKER. The latter is known

to have been used in the early 17th cent. as a term of opprobrium and also euphemistically altered to 'windsucker', so that the present bird name is conceivably as old as that. Comparable euphemistic formations are seen in the Flycatcher name CHERRY SUCKER and the Woodpecker name WOOD SUCKER.

Wind Throstle, Wind Thrush, see WHIN THRUSH.

(Wing. A loan from Old Norse **wængja* becoming **wengja* (cf. Norwegian *venga*) whence regularly Middle English *wenge* (disyllabic), from 14th cent. *winge*. This word, which supplanted two Old English terms *fiþere* and *feþra* (the latter simply the plural of *feþer* feather), occurs as the second element in several bird names, as REDWING. It goes without saying that such names cannot be older than the Middle English period. In LAPWING, the present word appears as the result of folk etymology, likewise in WING THRUSH.)

Wing Thrush, Winnard, see WHIN THRUSH.

Winter Crow. A term for the Hooded Crow occurring in Turner 1544 and taken up by Cotgrave 1611. It will have arisen in one of our eastern districts to which this Crow migrates in wintertime, the Carrion Crow being here the permanent resident.

Wipe. A name for the Lapwing from East Anglia and Lincs., and also Northumb. It descends from Old Norse *vípa*, the literal meaning of which is crest; cf. LAPWING.

Wirrangle, see WARIANGLE.

Witch. A seamen's name for the Stormy Petrel, first in Pennant 1785 'Stormy Petrel . . . hated by the sailors, who call them witches, imagining they forbode a storm'.

Wittol, see WHITE TAIL.

Witwall. Turner 1544 quotes 'Witwol', a contemporary German name for the

Golden Oriole, as though it were English as well, and the word appears with this meaning in 17th cent. sources beginning 1601 'Witwall or Loriot . . . is all ouer yellow'. At the same time, under the influence of genuine English WOODWALL, a (secondary) Woodpecker name, the foreign word was sometimes used in this sense, too. On German for English in Turner, see under SISKIN.

Woodchat, also **Woodchat Shrike.** The latter name introduced by Yarrell 1843, otherwise simply Woodchat, as in Pennant 1768 who found the term in Ray (posthumous) 1713; BOU 1883 'Woodchat', Hartert 1912 'Woodchat Shrike', now the usual form. It is inconceivable that this rare species ever acquired a popular English name, so that the primary source must be at fault. If the word Woodchat is genuine, and not the result of a printing or scribal error, then it must have designated some other bird.

Woodcock. Though we may well feel today, in the light of WOOD HEN, that the present name is properly applicable to the male bird, it has nonetheless been used since the earliest times to designate the species. It goes back via Middle English *wodecoc* to Old English *wuducocc*, known since *c*.1050. There are two Old English synonyms having the same literal meaning: *wuduhona* and *holthana* (*holt* wood, *hona* or *hana* cock; see HEN). The foregoing names contrast with yet another Old English synonym, the semantically more precise *wudusnīte* literally wood snite i.e. wood snipe. The last synonym will be the more original, the others, as their vagueness indicates, being replacements created by fowlers. Seeing that we have no less than three of them—indeed there may be four (see WOOD HEN) if not five (see SNITE)—it becomes evident that the original name was commonly avoided. In other words, name taboo will have been responsible for the neologisms—the Woodcock was, of course, a much

sought-after gamebird. See further under COCK, also BEAK.

Woodcock Pilot. A Yorks. name for the Goldcrest, said to precede the returning Woodcock by a couple of days.

Wood Culver. An obsolete name, recorded between *c*.1100 'wudeculfre' and 1688 'Ring-dove or Wood-culver', denoting the Wood Pigeon, and presumably also the Stock Dove, since the epithet *wood* vaguely denotes woodland birds, i.e. wild, as opposed to domesticated, species. Cf. WOOD PIGEON.

Wood Grouse. A term coined by Pennant 1776 'Wood Grous' in preference to the native (Scottish) name CAPERCAILLIE. The neologism was taken up by most of his successors, but Gould 1837 'Capercailzie' broke with this tradition, after which the present term was little used.

Wood Hack. A Lincs. name for the Green Woodpecker, first recorded in *Prompt.Parv. c*.1440 'Wodehake, or reyne fowle'.

Wood Hen. This little used term for the female Woodcock is nevertheless traditional: Middle English *wodehen*, Old English *wuduhenn* (Whitman 1899). Whether it originally denoted the female is, however, open to question. The sexes are alike; as the fowlers of long ago were never scientific ornithologists, they would hardly have needed a distinguishing term. It is therefore not unlikely that the present word was simply another evasive name for a much sought-after species; see under WOODCOCK.

Wood Knacker. A Woodpecker name from Hants., literally wood knocker.

Woodlark. The name first occurs *c*.1300–25 as a (doubtless makeshift) gloss on Old French *calaundre* calandra lark, next *c*.1340 'chalaundre and wodelarke'. From Turner 1544 'Wodlerck' onwards, the term clearly refers to the native species, the present spelling being current since Merrett 1667. The habits of the bird in question are not unlike those of the Skylark; it may, however, live on the edges of woods and perch on trees. The fowlers of long ago will have observed this difference, whence the name. It served to distinguish this species from the Skylark which, until the age of modern ornithology, was simply called LARK.

Wood Owl. A quite common name for the Tawny Owl, from the woodland habitat.

Woodpecker. First in Palsgrave 1530 'Woodpecker, *espec*' and well known to the early naturalists, Pennant 1768 effectively standardizing it. For a much older term, see SPEIGHT.

An exceptionally large number of Woodpecker names have developed locally, some having the same motivation as the present term, cf. WOOD HACK, WOOD KNACKER, WOOD SUCKER, also TAPPER. Many are associated with HICK-WALL. Not a few derive from the voice, see under YAFFINGALE. Others recall the bird's reputation as a weather prophet, e.g. RAIN BIRD. Reference is almost exclusively to the Green Woodpecker, at least in the first place; for an original term for a Spotted Woodpecker, see FRENCH PIE.

Wood Pigeon. Patterned on the older term WOOD CULVER, and in the first place similarly denoting either of our two woodland species. As a consequence, there was some overlapping with STOCK DOVE, as in the earliest attestations beginning with Merrett 1667 '*Oenas* . . . Stock Dove, or Wood Pigeon', and similarly in Charleton 1668 and Ray 1678. The later naturalists, however, from Pennant 1768 on, using STOCK DOVE only for *Columba oenas*, disregarded the present imprecise term. Their typical word for *C. palumbus* was RING DOVE. In spite of this, Wood Pigeon became in popular use more and more attached to this species, resulting in competition with the otherwise so well established and systematically

appropriate Ring Dove, until Hartert 1912 rather surprisingly made it the standard term, at least in ornithological work.

Wood Sandpiper. Coined by Pennant 1785 on the strength of information that, in its northern habitat, *Tringa glareola* may breed in swampy woodland. The bird has, of course, no traditional English name, indeed, it was not recognized as a British species at all until Montagu 1802. The name is not particularly apposite; it would, in fact, have better suited *T. ochropus* which commonly nests in trees, but Pennant had already coined GREEN SANDPIPER to name that bird.

Wood Speck. An East Anglian term for Woodpecker, recorded in the 16th and 17th cent., originally simply SPECK. Apparently replaced by a later variant **Wood Spack**, recorded by Halliwell 1847.

Wood Speight. An East Anglian and Glos. name for a Woodpecker, first in 1544 'Wodespecht', originally simply SPEIGHT. Also spelt **Wood Spite**.

Wood Sucker. A Woodpecker name from Hants., euphemistically for original *Wood Fucker literally wood beater, cf. WOOD KNACKER; original and euphemism are explained under WIND FUCKER, WIND SUCKER.

Woodwall. A Woodpecker name, in Somerset specifically the Green, in Hants. the Spotted. As a Woodpecker name, it is traceable to *c*.1489, the present spelling since 1566. But the name occurs in three much earlier attestations: *a*.1250, *a*.1310, and *c*.1325 'wodewale'. In the first two instances the context gives no indication of species, but in the third instance, the name glosses French *oriol* golden oriole, a meaning still found in the 16th cent. This is doubtless the primary sense, because the word has an exact correspondence in related languages on the Continent, where it is the oldest known name for the Golden Oriole, i.e. Dutch *wielewaal*, formerly *wiedewaal*, Low German *Wiedewol*, Middle High German *witewal*, this last the source of Turner 1544 'witwol', see WITWALL. Cf. HOOD AWL.

The various forms of this name presuppose a West Germanic prototype **widuwalō*, the first component *widu*-wood (in both senses), the second conceivably based on a syllable *wal* echoing the distinctive call of the male—we recall that this furtive denizen of the treetops is more often heard than seen, and compare the etymology of ORIOLE.

One may ask how a West Germanic name could survive in English for a bird so little known in this country as the Golden Oriole. We must infer that it was formerly a relatively abundant species, indeed until the 15th or 16th cent. Then it became rare and the name was transferred to the Green Woodpecker, the meaning found in Somerset. Both species are closely associated with life in the trees, and the colour of the female Oriole is predominantly green. Reference in Hants. to a Spotted Woodpecker must be secondary, names proper to the Green quite often being shifted to the Spotted. We note, in conclusion, that Woodwall is found only in the far south, as expected of a name which used to designate the Golden Oriole.

Wood Warbler. Coined by Yarrell 1843 for systematic reasons in place of the earlier folk-name WOOD WREN.

Wood Wren. A local name, known from Somerset, introduced into ornithological literature in 1794 (*OED*), and the usual term in books until Yarrell 1843 converted it into WOOD WARBLER. The epithet *wood* refers, of course, to the favoured woodland habitat.

Woosell, see OUZEL.

Wran. In common use in local speech in Scotland, it goes back via Middle English *wranne* to Old English *wrænna*, an irregular variant of *wrenna*; see WREN. This form also survives at the other end of the country in Cornwall. The pet form **Wrannie** is common in Scotland, also **Wrannock**, with the diminutive suffixes -IE, -OCK.

Wren. A traditional name, Middle English *wrenne*, Old English *wrenna*, implying a West Germanic **wrannjan*. An older variant **wrandjan* is presupposed by Old High German *wrendo*; this last forms a diminutive *wrendilo*, comparable to Old Norse *rindill*. The basic sense is (little) tail, a reference to the perky, cocked-up tail, unique among our birds and thus calculated to inspire a name. The same part for whole principle again in the synonyms CUT, SCUT, STAG, TOPE, also BOBBY WREN. Cf. WRAN.

Not surprisingly, the name of the Wren is often prefixed by an adjective meaning tiny or the like, as CHITTY WREN, PUGGY WREN, TITTY WREN. Or the name itself may denote smallness, e.g. CRACKY, TWO FINGERS. In popular use the name Wren is often transferred to other small birds, as MUFFIE WREN, REED WREN, WILLOW WREN, WOOD WREN.
Ams.St. 197–200 (1981)

Writing Lark. A name for the Yellowhammer, found here and there from Yorks. to Surrey, motivated by the lines of 'writing' on the egg-shell. In Ches., the name may locally become **Writing Master.** Cf. SCRIBBLING LARK.

Wryneck. Known since 1585, the name used by Merrett 1667 and Ray 1678, further Pennant 1768, making it the undisputed standard. *Wry-* meaning twisted refers to the habit of twisting the head right round on the neck. Other names for this once quite common species include BARLEY BIRD, CUCKOO'S MARROW, EMMET HUNTER, PEE or PEEL BIRD, RIND BIRD.

Y

(-y, see -IE.)

Yaffingale. A Green Woodpecker name in local use from Berks. to Somerset, with such variants as Berks. **Yelpingale**, Somerset **Yappingale**, the latter first in 1601 as 'Yippingale'. The term has occasionally appeared in literature, as Tennyson's 'garnet-headed yaffingale' (*OED*). The name is based on onomatopoeic syllables, i.e. *yaff* bark, *yelp, yap,* in imitation of the call, for the rest modelled, humorously we suppose, on NIGHTINGALE. Cf. YAFFLE; for further examples of voice motivation, see GALLEY BIRD, LAUGHING BIRD, NICKER, also HIGH HOE.

Yaffle, Yaffler. The name Yaffle, in use from Surrey to Worcs. to denote the Green Woodpecker, imitates the call, whence Heref. Yaffler. Cf. YAFFINGALE.

Yaldrin, see YOLDRING.

Yappingale, see YAFFINGALE.

Yarlin, see YORLING.

Yarwhelp, Yarwip, see GODWIT.

Yeldrin, see YOLDRING.

Yeldrock. A Northumb. name for the Yellowhammer, formed from YELDRIN by suffix change; another change in Ulster **Yellow Yeldrick**; see -ICK, -ING, -OCK.

Yell Am-bird, see YELLOW HAM.

Yellow Ammer, see YELLOWHAMMER.

Yellow An-bird, see YELLOW HAM.

Yellow Bunting. A term for the Yellowhammer created by Pennant 1776 (replacing 1768 'Yellow Hammer'), generic BUNTING corresponding to Linnaeus 1758 *Emberiza*. Pennant's coining was largely ignored until re-introduced by Yarrell 1843 'Yellow Bunting or Yellow Ammer', the former being treated as the main term. It maintained itself to some extent in later ornithological literature, but 'Yellow Hammer' was used in BOU 1883. Even though Hartert 1912 brought back the present term, it could not compete with its rival, now standardized as YELLOWHAMMER, always a well-known, living word, which has come into its own in spite of the systematic suitability of Yellow Bunting.

Yellow Ham. An alternation of YELLOW-HAMMER by folk-etymological substitution of *ham* for *hammer* occurring in Turner 1544 'yelow ham, yowlryng'. This unfortunate name was further doctored in 1657 'yellow-ham bird', surviving in Kent as **Yellow An-bird**, in Sussex as **Yell Am-bird**.

Yellowhammer. First recorded in 1587, but demonstrably somewhat older; see YELLOW HAM. It is found in Merrett 1667 and reported by Ray 1678 'yellow hammer, or amber', whence Pennant 1768 'Yellow Hammer'. Although replaced in Pennant 1776 by 'Yellow Bunting', the present name was used by his successors until Yarrell 1843 who, while re-introducing 'Yellow Bunting' as his main term (see further under this name), also admitted 'Yellow Ammer', but strictured the variant with 'Hammer'. However, we find BOU 1883 'Yellow Hammer'. Nevertheless it was not until the early part of this century that the present name established itself in ornithological literature as the preferred, or now usual, term.

Yellowhammer is a corruption of **Yellow Ammer**, best known in the literature from Yarrell (above). The earliest notice is Ray, above ('amber' with intrusive *b* for more original ammer, cf. GOLDEN AMBER). A contracted form is seen in 1656 'Yelamber', the same again in 1556 'yelambre'. The ephithet *yellow*, motivated by the resplendent head and breast of the male bird, is a secondary accretion. The oldest form of the name, found in Old English, is simply *amer* (manuscript corruptly 'amore') just like the German cognate *Ammer*. The name is of West Germanic age. A variant form of *yellow* lies behind YOULRING and its multifarious progeny.

Locally, the Yellowhammer may go under the name of GOLDFINCH.

Yellow Molly. A pet name for the Yellow Wagtail, heard in Hants.

Yellow Owl. A name for the Barn Owl, locality unspecified; it derives from the brownish-yellow upper parts. Cf. WHITE OWL.

Yellow Plover. A Border Country name for the Golden Plover.

Yellow Wagtail. A term known since Charleton 1668 and almost universally used; locally, it may include the Grey Wagtail. See YELLOW MOLLY.

Yellow Yeldrick, see YELDROCK.

Yellow Yite, see YITE.

Yellow Yoldring, see YOLDRING.

Yellow Yorling, see YORLING.

Yelper. A Lincs. name for the Avocet, motivated by the alarm call, here likened to yelping. Cf. BARKER.

Yelpingale, see YAFFINGALE.

Yerlin, see YORLING.

Yewlet, see HOWLET.

Yite. A Scottish name for the Yellowhammer, first recorded in 1812; it is imitative of the twittering call note. Commonly also **Yellow Yite.**

Yockle, also **Yuckle.** Names for the Green Woodpecker from Shrops. and Wilts. respectively, to all appearances echoing the call, likened to laughing (cf. LAUGHING BIRD), but perhaps originally associated with ICKLE, etc.

Yogle. A Shetland name for an Owl, from Norn, the ultimate source being Old Norse *ugla*, etymologically identical with OWL. Cf. KATOGLE.

Yoldring. A northern and Scottish name for the Yellowhammer, first in 1790 'Yold-ring', commonly becoming **Yoldrin** (see -ING). It is a regular development from **Youldring,** a more recently attested, but linguistically more archaic form. This, in turn, arose from YOULRING, with *d* appearing as a glide consonant. The present name has a number of secondary variants, notably **Yaldrin** and **Yeldrin,** see also YELDROCK.

The epithet *yellow* is often used with the above forms, e.g. **Yellow Yoldring,** the ultimate origin of the word (see YOULRING) being, of course, no longer recognized. Cf. YORLING.

Yoolet, see HOWLET.

Yorling, earlier **Yourling.** First in 1679 'Yellow-hammer or Yourling', developing from YOULRING, the consonants *l* and *r* changing places (metathesis) to bring the word into line with the many other bird names ending in -LING. Subsequently, the first syllable was regularly reduced to *yor-* and the second commonly to -LIN (see -ING), hence **Yorlin,** known since 1789. The name is in northern and Scottish use, also with secondary variants in **Yarlin, Yerlin.**

The epithet *yellow* is often used with the above forms, e.g. **Yellow Yorling,** the ultimate origin of the word (see YOULRING) being, of course, no longer recognized. Cf. YOLDRING.

Youldring, see YOLDRING.

Youlring, older spelling **Yowlring.** A northern and Scottish term for the Yellowhammer, current in the 16th cent. It first occurs in Turner 1544 'yelow ham, yowlryng', but in a linguistically more archaic form in 1571 'yowlorings', in normalized spelling (sg.) **yowlowring,* where *yowlow-* is a common variant of *yelow* emerging in the Middle English period.

The element -*ring* makes no sense in a Yellowhammer name. The original word must have been *yowlowling, i.e. with the name-forming suffix -LING added to the adjective (exactly as in the synonym BLAKELING), subsequently altered to -*ring* to avoid the two *l* sounds so near together (dissimilation).

In later language, the present term evolved in two different directions, giving rise to the divergent types exemplified by YOLDRING and YORLING.

Yourling, see YORLING.

Yuckle, see YOCKLE.